香港住宅通用设计指南

Universal Design Guidebook

For Residential Development In Hong Kong

香港房屋协会 编著

中国建筑工业出版社
CHINA ARCHITECTURE & BUILDING PRESS

著作权合同登记图字：01-2009-3780号
图书在版编目（CIP）数据

香港住宅通用设计指南／香港房屋协会编著.— 北京：中国建筑工业出版社，2009
ISBN 978-7-112-11142-8

I. 香… II. 香… III. 住宅—建筑设计—香港—指南
IV. TU241-62

中国版本图书馆CIP数据核字（2009）第119351号

本书乃香港房屋协会根据香港的住宅情况编订，与中国大陆之住宅情况或有不同之处，请读者自行判断是否适用于中国大陆之住宅建筑，出版社、著作权人及参与制作本书之所有顾问人员和机构，均不对因应用相关设计而产生的任何不利后果承担责任。

责任编辑：戴　静　丁　夏

香港住宅通用设计指南
Universal Design Guidebook For Residential Development In Hong Kong
香港房屋协会　编著
*
中国建筑工业出版社出版、发行（北京西郊百万庄）
各地新华书店、建筑书店经销
利丰雅高印刷（深圳）有限公司制版
利丰雅高印刷（深圳）有限公司印刷
*
开本：880×1230毫米　1/16　印张：9½　字数：310千字
2009年9月第一版　2009年9月第一次印刷
定价：98.00元
ISBN 978-7-112-11142-8
(18373)

总咨询顾问
利安顾问有限公司

LEAD CONSULTANT
Leigh & Orange Limited

香港房屋协会
Hong Kong Housing Society
29/F World Trade Centre,
280 Gloucester Road,
Causeway Bay, Hong Kong
香港铜锣湾告士打道280号
世界贸易中心29字楼
Tel 电话: (+852) 2839 7888
Fax 传真: (+852) 2882 2001
Website网址: www.hkhs.com

顾问

医疗、卫生及复康顾问
林正财医生 铜紫荆星章 太平绅士
基督教灵实协会

畅达设计顾问
关国乐先生 荣誉勋章
生活环境辅导服务
复康资源协会

通用设计顾问
廖维武先生
香港中文大学

职业治疗顾问
郑吴倩华女士
张秀娴女士
卢景笙先生
陈国强先生
香港职业治疗学会

ADVISORS

Medical, Health and Rehabilitation Advisor
Dr. Ching Choi Lam, BBS, JP
Haven of Hope Christian Service

Accessibility Advisor
Mr. Joseph Kwan, MH
Environmental Advisory Service
Rehabaid Society

Universal Design Advisor
Mr. Laurence Liauw Wie Wu
Chinese University of Hong Kong

Occupational Therapy Advisors
Ms. Serena Cheng
Ms. Sau Han Cheung
Mr. Eddy Lo King-sang
Mr. Simon Chan
Hong Kong Occupational Therapy Association

Acknowledgements

Thanks are offered to the following for their support, their attendance of the Universal Design Workshop and their valuable comments.

谨此感谢以下各界参加通用设计研讨会，并且提供宝贵意见。

鸣谢

- Caritas Hong Kong
- Direction Association for the Handicapped
- Hong Kong Blind Union
- Hong Kong Council of Social Services
- Hong Kong Family Welfare Society
- Hong Kong SAR Government, Buildings Department
- Hong Kong SAR Government, Food and Health Bureau
- Hong Kong SAR Government, Housing Department
- Hong Kong SAR Government, Planning Department
- Hong Kong SAR Government, Social Welfare Department
- Hong Kong Polytechnic University, School of Nursing
- Hong Kong University, Faculty of Architecture
- The Hong Kong Federation of Youth Groups
- The Hong Kong Institution of Engineers
- The Hong Kong Institute of Surveyors
- Hong Kong PHAB Association
- Hong Kong Rehabilitation Power
- Hong Kong Society for the Deaf
- The Hong Kong Society for Rehabilitation
- Playright Children's Play Association
- Rehabilitation Alliance Hong Kong
- Retina Hong Kong
- The Salvation Army
- Senior Citizen Home Safety Association
- Tung Wah Group of Hospitals

[The above list is in alphabetical order]

- 香港明爱
- 路向四肢伤残人士协会
- 香港失明人协进会
- 香港社会服务联会
- 香港家庭福利会
- 香港特别行政区政府屋宇署
- 香港特别行政区政府食物及卫生局
- 香港特别行政区政府房屋署
- 香港特别行政区政府规划署
- 香港特别行政区政府社会福利署
- 香港理工大学护理学院
- 香港大学建筑系
- 香港青年协会
- 香港工程师学会
- 香港测量师学会
- 香港伤健协会
- 香港复康力量
- 香港聋人福利促进会
- 香港复康会
- 智乐儿童游乐协会
- 香港復康联盟
- 香港视网膜病变协会
- 救世军
- 长者安居服务协会
- 东华三院

[以机构英文名称字母顺序排名]

Contents

目录

Contents

Chapter 3 Barrier-Free Access

目录

第3章 无障碍设计

Contents

Chapter 3 Barrier-Free Access

目录

第3章 无障碍设计

Contents

目录

第4章　家居安全

第5章　环境因素

Contents

Chapter 6 Good Practices on Detailing

目录

第6章　装饰详细设计的推荐方案

Contents

Chapter 7 Renovation and Conversion

目录

第7章 整修和改建

章 Cha

1

pter

Introduction

引言

Introduction

1

1.1

1.1 BACKGROUND

Universal Design is an emerging approach which aims to develop a theoretical framework and practical means to a more user-friendly living environment. It aims to do this specifically by incorporating features accommodating the widest possible range of human needs. In doing this, it has a close relationship with Accessibility Design, yet it is important to differentiate between the two. Accessibility design necessarily focuses on the sometimes specialised needs of groups whose mobility has traditionally been limited. Universal Design takes that body of knowledge and seeks to extract from it, features which can become the norm, for the benefit not only of special groups but of everyone.

Universal Design incorporates accessibility issues but seeks a broader consensus as a means to broader application. In that sense, Universal Design needs to respect the average and to take account of related issues, such as cost effectiveness, space efficiency, flexibility and identity. At its most effective, Universal Design can be invisible. It is an inclusive approach which embodies the hope that the largest number of people can access the widest variety of places with the greatest degree of safety and convenience. It is important to all of us because we are all challenged at various times of our lives. We were all children. We will all become ill. If we are lucky, we will become old. By some definitions, a large proportion of society is in some way disabled; even if people prefer not to think of themselves, or to present themselves in that way. In many societies with an aging population, that proportion will rise, as the number of caregivers declines.

Many countries have legislated on behalf of the aged and people with disabilities and have produced design guidelines, particularly related to accessibility design over a number of decades. The contributors to this Guidebook think that Universal Design is the next step.

We have looked closely at research and legislation around the world before attempting to identify features most relevant to Hong Kong. That is a particular challenge because this is one of the most demanding places in the world for which to design. Nowhere else is space used more efficiently. We want to make Hong Kong buildings even more efficient; by making them more accessible, safer and more convenient to use, particularly by people with specialist needs but also by all of us. This document is a starting point for a process of exploration which stretches ahead.

1.1 背景

通用设计作为一个崭新的设计手段，渐渐出现于当代设计潮流中。它将设计理论构架与实践相结合，使设计元素能尽可能地满足广泛人士的不同需求，以创造出一个方便用户使用的生活环境。为达到此目的，通用设计与无障碍设计（Accessibility Design）有紧密的联系，然而二者却有本质上的区别。畅达设计着重顾及传统概念上身体活动受到限制的人士的特殊需求，而通用设计取畅达设计概念之精华，并使可普遍使用的元素成为设计的标准，不仅让有特殊需要的人士，而且让每一个人都可受益。

通用设计包含畅达设计，但它亦寻求更广泛意义上的一致性，以在更广范围内得以应用。因此，通用设计需要尊重有一般需求的人，并考虑与其相关的问题，例如成本效益、空间效用、灵活性和同一性。最有效的通用设计可以是无形的。它是一种无歧视的设计途径，并体现了一种希望，即让尽可能多的人士能以最安全和最方便的途径，到达不同类型的地方。这对我们每一个人都很重要。因为我们在人生的不同阶段都会面临不同的挑战。我们都曾是小孩。我们都会生病。若幸运的话，我们都会有终老的一天。根据有关定义，社会上大部分人士都有各种程度的障碍，即使他们不愿视自己为残障人士，或不愿如残障人士般生活。在很多人口老龄化的社会，由于护理人士数目相应减少，此类残障人士的比例将会增加。

数十年来，许多国家代表长者和残障人士立法，并制定了畅达设计指南。本书认为，通用设计将会成为未来方向。

在尝试确认适合香港的通用设计标准之前，我们仔细查阅了世界各地相关的研究和法例。香港是全球在设计方面要求最多的地区之一，并且十分重视有效利用空间。因此，提出适合香港情况的通用设计标准有极大的挑战性。我们希望利用通用设计，使香港建筑物更加畅达和安全，不仅满足有特殊需要的人士的需求，而且方便每一个人使用，从而使香港建筑物更具空间及经济效益。本书是此长远探索过程的起步点。

引言

1.2 通用设计定义

Ron Mace（梅斯）在1988年提出一个国际认可的通用设计定义:

「通用设计是一种设计途径，它集合了能在最大程度上适合每一个人使用的产品及建筑元素。」

[Preiser, 2001]

1.2 DEFINITION OF UNIVERSAL DESIGN

Ron Mace's definition of Universal Design in 1988 has been recognized internationally:

"Universal Design is an approach to design that incorporates products as well as building features which, to the greatest extent possible, can be used by everyone."

[Preiser, 2001]

1.3 通用设计原则

于1997年，美国北卡罗来纳州州立大学的通用设计中心有一群研究员制订了以下七项通用设计原则:

原则一：平等使用
设计对于不同能力人士都是可以使用和值得购买的。

原则二：方便灵活
该设计适应广泛的个人偏好和个人能力。

原则三：简单易用
该设计简单易用，不会因使用者的经验、知识、语言技能或当下精神集中程度而有异。

原则四：资讯简明
该设计有效地传达所需要的资讯，不会因环境或使用者感知能力而有异。

原则五：容差纳误
设计使危险及事故和误操作产生的危害和不良影响最小化。

原则六：省力易用
可以有效地和舒适地使用该设计，省力而不倦。

原则七：可用面积和空间
为各种体型、姿态和活动能力的使用者设计合适的接近、到达、操作和使用的尺寸和空间。

[Preiser, 2001]

这七个原则说起来简单，但实践起来复杂难行。本指南探索实际的方法，应用于香港的住宅发展。

1.3 PRINCIPLES OF UNIVERSAL DESIGN

In 1997, a group of researchers of the Centre for Universal Design, NC State University of United States, developed the following seven principles of Universal Design:

Principle One: Equitable Use
The design is useful and marketable to people with diverse abilities.

Principle Two: Flexibility Use
The design accommodates a wide range of individual preferences and abilities.

Principle Three: Simple and Intuitive Use
Use of the design is easy to understand, regardless of the user's experience, knowledge, language skills, or current concentration level.

Principle Four: Perceptible Information
The design communicates necessary information effectively to the user, regardless of ambient conditions or the user's sensory abilities.

Principle Five: Tolerance for Error
The design minimizes hazards and the adverse consequences of accidental or unintended actions.

Principle Six Low Physical Effort
The design can be used efficiently and comfortably and with a minimum of fatigue.

Principle Seven: Size and Space for Approach and Use
Appropriate size and space is provided for approach, reach, manipulation, and use regardless of user's body size, posture, or mobility.

[Preiser, 2001]

The seven principles are simple in words but complex to attain. This Guidebook explores practical ways to apply them to Hong Kong residential development.

Introduction

1.4 UNIVERSAL DESIGN PYRAMID

1.4

Different groups can have different needs. Universal Design aims for appropriate balance with considerations for the group of people who have the greatest need for care. The following pyramid describes the hierarchy of different groups' needs:

In the above pyramid, Group A, including Row 1 and 2, are able-bodied people. Current architecture design aims for a full consideration of this group. Group B including rows 3, 4 and 5, are people who need caregivers or who have disabilities. They need more design features than Group A. Group C, namely row 6, are people in wheelchairs able to live independently. People in wheelchairs have higher requirements of spatial design. Group D includes people in wheelchairs who need caregivers. In this Guidebook, more efforts will be taken for the residents of Group B, C and D, whose special needs are not fully provided for by current architecture practice in Hong Kong.

1.4 通用设计金字塔

不同组别人士有不同需要，通用设计的目的是在照顾最有特别需要的人士同时，提供适当的平衡，以满足不同组别人士的需要。以下的金字塔显示各组别需要的层次：

以上金字塔的A组（即第1和第2行）是健全人士，现有的建筑设计完全考虑到这一组别的需要。B组（即第3、第4和第5行）是需要别人照顾的人或残障人士，这组别比A组需要更多的特殊设计元素。C组（即第6行）是可以独立生活的轮椅人士，在空间设计上有较高的要求。D组（即第7和第8行）是需要别人照顾的轮椅人士。香港现时的建筑设计未全面顾及B组、C组和D组人士的特别需要，因此本指南会着重讨论。

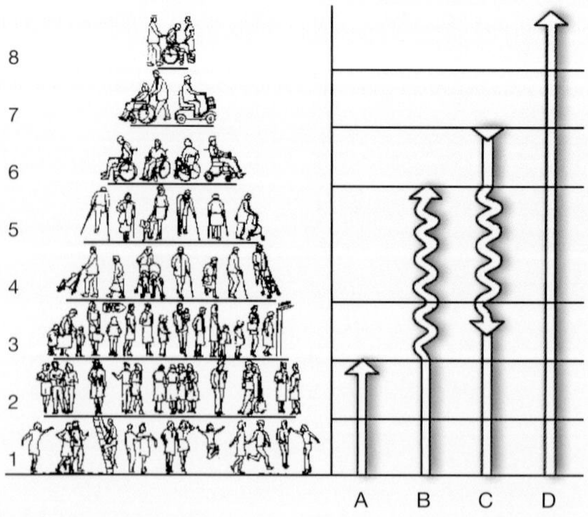

Universal design pyramid
通用设计金字塔

[Preiser, 2001]

1

1.5 目的与范畴

本指南有关通用设计的目的和范畴，可作如下定义：

通用设计是一种崭新的设计途径，创造可供不同组别社群使用的环境，包括婴儿、幼儿、孩童、青年、成年人、长者、残障人士、体弱者和孕妇，其设计安全、方便，具灵活性并可变更用途。此设计途径必须先了解和体谅使用者在一生中不同阶段的各种需要和能力的改变。

本指南的目的是为住宅发展提供通用设计的指南，以切合最大部分居民的需要。当中必须了解到居民的能力会因年龄、残障和疾病而改变；除了平等使用外，此指南也考虑到灵活性、简易性、舒适、健康、安全、方便、畅达、实用、容易维修、耐用和可持续性等因素。

编写本指南的目的如下：

- 为建筑师和设计师介绍设计策略和推荐方案；
- 提出可持续住宅发展规划的指引；
- 协助发展商明白各种组别居民的需要吸引最大的市场，和更切合居民对建筑和管理的需求；
- 为居民提供整修、使用和管理策略的建议，帮助他们改善居住环境；
- 在相关的公众地方和公众设施推广通用设计的原则。

1.5

1.5 OBJECTIVES AND SCOPE

The objectives and scope of Universal Design for this Guidebook can be defined as follows:

Universal Design is an emerging approach to create environments which are usable by the whole spectrum of community, including infants, toddlers, children, the young, the average, the elderly, disabled, weak, and pregnant, with safety, convenience, flexibility and convertibility. It is a design approach which requires understanding and empathy for the range of human needs and abilities throughout life.

This Guidebook aims to provide Universal Design guidelines for residential development intended to cater for the broadest range of residents. It requires an understanding of varying abilities with age, disability and illness. In addition to equitable use, it considers flexibility, simplicity, comfort, health, safety, convenience, accessibility, practicality, ease of maintenance, durability and sustainability.

This Guidebook has the following objectives:

- To introduce design strategies and good practice for architects and designers;
- To provide guidelines for sustainable residential development planning;
- To help developers identify the needs of different groups of residents in order to appeal to the broadest market and better meet the needs of residents in terms of both of building and management;
- To suggest renovation, utilisation and management strategies for residents to improve their living environment;
- To promote Universal Design principles for related public areas and provisions

Introduction

1.5

The Hong Kong Government has made provisions with regards to accessibility and development planning in statutory documents such as the Disability Discrimination Ordinance, 1996, the Residential Care Homes (Elderly Persons) Ordinance and Hong Kong Planning Standards and Guidelines (HKPSG). Subsequently, the Design Manual - Barrier Free Access 1997 was published by the Buildings Department, as a design standard. To enhance this Design Manual, the Buildings Department has commissioned a consultancy study and the revised Manual has been formulated.

In addition, this Guidebook will introduce good practice and guidelines for Universal Design. It also aims to draw public attention to the issues of universal design for residential development, and hence help to improve the living environment in Hong Kong.

香港特区政府在《残疾歧视条例1996》、《安老院条例》和《香港规划标准与准则》等法定文件中，设定了关于畅达设计和发展规划的法律条文。此后屋宇署在1997年出版的《设计手册：畅通无阻的通道1997》是无障碍设计的规范条文。为加强此设计手册的内容，一个顾问研究小组受屋宇署委托，制订了此手册的修订版。

在此基础上，本指南将引进通用设计的推荐方案和设计指南，目的是吸引公众关注住宅发展的通用设计问题，从而协助改善香港的居住环境。

1

1.6 纲要

本指南共有七章，分列如下：

第1章 引言

第1章介绍通用设计的理论，阐明本指南的范畴和目的。

第2章 规划和空间设计指南

第2章提出有关通用设计的城市规划问题和讨论不同组别居民对空间的要求，包括年轻人、长者、残障和体弱人士，并且举例说明适用于香港的通用设计。

第3章 无障碍设计

第3章介绍无障碍设施的设计，涉及住宅使用空间的规划和设计，及住宅内交通设施设计。

第4章 家居安全

第4章介绍无障碍疏散设计，并阐述伤害预防和家居安全的策略，以满足各组别居民的需求。

第5章 环境因素

第5章说明通用设计与优质生活的关系，认为促进健康的元素包括室内空气质量、温度和视听环境。

第6章 装修详细设计的推荐方案

第6章介绍扶手、装修物料和其他有关通用设计的建筑详细设计。

第7章 整修和改建

第7章把通用设计应用于现有楼宇，并为发展商和居民提供整修和改建策略。

1.6

1.6 FRAMEWORK

There are seven sections to this Guidebook, as follows:

Chapter 1 Introduction

Chapter One introduces the theory of Universal Design. It also provides the scope and objectives of this Guidebook.

Chapter 2 Planning and Spatial Standards

Chapter Two introduces urban planning issues related to Universal Design and discusses the spatial needs of particular residents, including the young, elderly, disabled, and weak. It provides examples of Universal Design, applicable for use in Hong Kong.

Chapter 3 Barrier-Free Access

Chapter Three introduces barrier-free provisions. It considers the planning and design of residential spaces and associated circulation areas.

Chapter 4 Home Safety

Chapter Four introduces standards for means of escape. It describes injury prevention and home safety strategies for different groups of residents.

Chapter 5 Environmental Factors

Chapter Five relates Universal Design to quality of life. It considers wellness in relation to in-door air quality, thermal, visual and acoustic considerations.

Chapter 6 Good Practice Detailing

Chapter Six provides architectural details for handrails, finishes and other factors related to Universal Design issues.

Chapter 7 Renovation and Conversion

Chapter Seven applies Universal Design to existing buildings, and introduce renovation and conversion strategies to developers and residents.

章

Cha

pter 2

Planning and Spatial Design Guidelines

规划和空间设计指南

Planning and Spatial Design

Universal Design in housing is an approach to sustainable residential development planning. It focuses not only on accessibility, but also on social inclusion issues such as health, education, entertainment, communication, care and safety. To achieve this, it is important to understand the needs of all residents. Therefore, the needs of different individual groups will be addressed in the first part of this Chapter.

The second part focuses on flat layouts which can be applied in Hong Kong, with considerations for diverse abilities.

房屋的通用设计是达到可持续住宅发展规划的一个途径，不但注重畅达性，也关注健康、教育、文娱、通讯、护理及安全等社会问题。为达到此目标，在设计上必须明白不同组别居民的需求。因此，本章的第一部分首先论述不同组别居民的需要。

第二部分侧重于介绍可在香港推行又顾及不同人士能力的住宅单位平面设计。

2.1 NEEDS OF DIFFERENT GROUPS

Different social groups have different requirements of residential development planning. Provision of proper community facilities is an essential means to meet the needs of different groups of residents. To identify appropriate provision, research has been carried out by Hong Kong SAR Government, and the Planning Department has incorporated the findings in the Hong Kong Planning Standards and Guidelines [HKPSG]. The HKPSG recommends to provide community facilities based upon the growth or concentration of population in a given area. The guidelines are useful to assess land requirements for community uses. As the number and age distribution of residents and hence demand for facilities will change, it is desirable that assessment of community facilities shall be carried out before a residential development is planned, and repeated regularly after the development is constructed. The following items should be included for development assessments.

- Current age distribution for existing neighbouring residential developments;
- Predicted age distribution of residents for the proposed residential development;
- Current modes of transportation and their impact on the proposed residential development and its community facilities;
- Predicted modes of transportation in the coming decades;
- Existing neighbouring facilities available to new residents of the proposed residential development;
- Flexibility of existing neighbouring facilities for functional conversion.

It is recommended that community facilities shall be functionally convertible, and more concerns shall be taken for people with special needs. As abilities vary with age, a good understanding of the needs of different groups of residents is essential to generate planning solutions.

2.1 不同组别的需要

不同的社会组别居民对住宅发展规划有不同的要求，我们必须提供合适的社区设施，以满足不同组别居民的需要。为了提供合适的设施，香港特区政府已进行研究，并由规划署根据研究结果制订《香港规划标准与准则》，建议根据某一地区的人口增长和集中程度，提供社区设施。上述规划标准有助评估作为社区用途的土地需求。由于居民的数目和年龄分布会改变对设施的需求，有关方面应在进行住房发展规划前，对社区设施进行评估，并在发展完成后定期进行再评估。下列各项应包括在发展评估项目中：

- 现时邻近住宅区的居民年龄分布；
- 预计规划的住宅区中，居民的年龄分布；
- 现时的交通模式及其对规划的住宅区和社区设施的影响；
- 预计未来数十年的交通模式；
- 可供规划的住宅区的新居民享用的现有邻近设施；
- 现有邻近设施进行功能改建的灵活性。

建议社区设施可在功能上作出改建，并更应关注有特别需要的人士。由于居民的能力会随着年龄的增长而改变，详细了解各个居民组别的需要，是制订规划方案的关键。

Guidelines

规划和空间设计指南

2.1

2.1.1 Children

Nursing and Care

- It is preferable that nursery schools and kindergartens are available in the neighbouring areas of residential buildings.
- It is preferable that healthcare facilities, such as children's clinic are available in the neighbouring areas of residential buildings. This offers convenience, promotes neighbourhood identity and encourages prompt attention to health problems.

Education

- It is preferable that school facilities shall be available near residential developments. This offers convenience, promotes neighbourhood identity and encourages safety for children.
- In areas of heavy traffic, it is preferable that entrances of school facilities shall be located away from main civic roads and that waiting zones shall be provided nearby for parents and caregivers. Entrances and waiting areas shall be clearly visible.
- It is preferable that cultural facilities, such as libraries and bookstores, shall be available near residential developments.

2.1.1 儿童

看护和护理

- 幼儿园及幼稚园最好设于住宅的邻近范围。
- 医护设施，如儿童诊所等，最好设于住宅的邻近范围，以方便居民、促进邻里活动，并鼓励居民及时关注健康问题。

教育

- 学校最好设于住宅区附近，以方便居民、促进邻里活动，并有利于为儿童提供安全环境。
- 在交通繁忙的范围，学校入口最好远离主要交通道路，并须在入口附近为父母及看护者设置等候区。入口及等候区应该清晰易见。
- 文化设施，例如图书馆及书店等，最好设于住宅区附近。

Kindergarten in neighbouring areas of residential buildings
住宅附近的幼稚园

Children's Play Areas

- It is recommended that an outdoor play areas with accessible and challenging play settings shall be provided near residential developments. Where resources are available and likely levels of use justify it, it is preferable to provide age-appropriateness options with opportunities for motor skill development, decision-making, learning, fantasy play, and social development.
- It is preferable that play settings shall provide access to sunshine, shade and protection from adverse conditions.

Good Practice

- It is preferable that an open space with planting shall be provided adjacent to the outdoor play area. Flowers and plants should be carefully selected to enhance children's interaction with the natural environment.
- It is preferable that cycling spaces shall be provided and integrated into the residential developments with safety considerations.
- It is preferable that indoor play areas shall be provided where resources are available and likely levels of use justify it.

儿童游玩场所

- 建议在住宅区附近设置方便易达及有趣味的户外游玩设施。在资源供应允许及使用合理的条件下，最好应提供适合不同年龄的游玩设施，让儿童借此机会发展肌能技巧和决策能力，并有机会学习、游玩并提高社交能力。
- 游玩设施的地点最好有阳光照耀，同时设有遮荫之处，使儿童不受恶劣天气所影响。

推荐方案

- 于户外游玩场地旁，最好种植花草树木，并应小心选择花朵及植物，以提高儿童与大自然的互动。
- 在顾及安全的条件下，住宅区内最好设置自行车道及场地。
- 在资源供应允许及使用合理的情况下，最好附设室内游玩场所。

Children's play settings in neighbouring areas of residential buildings
住宅附近的儿童游玩设施

规划和空间设计指南

安全

预防儿童受伤受到广泛关注。近期研究显示在香港家居意外中，年龄介乎5至9岁的儿童很容易跌伤及受到其他与活动有关的意外。要避免儿童受伤，住宅发展规划应注意下列事项：

- 最好设置确保儿童安全及低危险的游玩设施。
- 建议于游玩场所内，为家长及看护者设监护点。
- 在交通繁忙的范围及资源许可下，最好将车辆通道、自行车道及行人路予以分隔。
- 如有足够资源而又合乎实用原则，最好为不同年龄组别的儿童提供不同的游玩场所。
- 最好为不同年龄组别的儿童提供相应的游玩设施。

推荐方案

- 户外游乐场需维修良好，并定期由受训的管理人员检查，以确保设备的安全性。
- 建议在游玩场地的入口，设置使用者年龄范围及安全游玩的提示牌。

Safety

Prevention of injury to children is of wide concern. A recent study in domestic injuries in Hong Kong demonstrated that children at the age from 5 to 9 are susceptible to falls and other motor-related injuries. To prevent injury, it is recommended that residential development planning shall consider the following issues:

- It is preferable that child-proof and hazard-free play settings shall be provided.
- It is recommended that play areas shall facilitate supervision by parents or caregivers.
- In areas of high traffic and where resources allow, it is preferable that vehicular, bicycle and pedestrian systems shall be separated.
- Where resources allow and levels of usage justify it, it is preferable that separate play areas for different age groups shall be provided.
- It is preferable that play equipment shall be provided for different age groups.

Good Practice

- Outdoor playgrounds need to be well maintained and regularly inspected by trained management staff to ensure a non-hazard provision.
- It is recommended that user age range and cautionary signage for play safety shall be provided at the entrances of play areas.

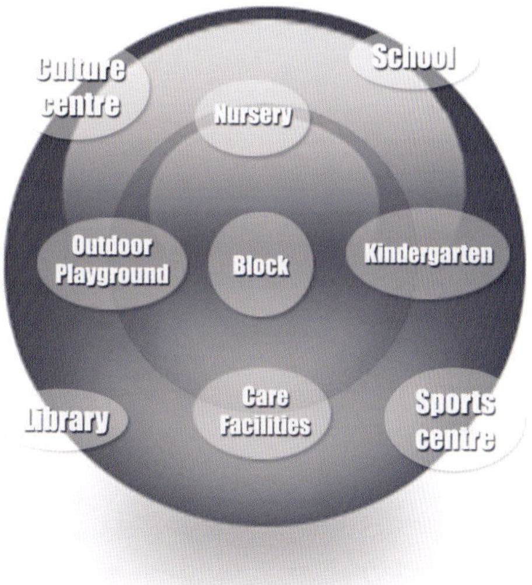

Sport facilities in the neighbouring areas of residential buildings
住宅附近的体育设施

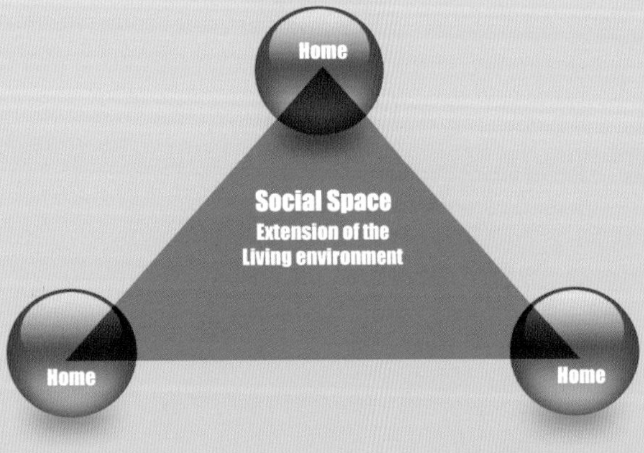

2.1

2.1.2 Youth

Youth Recreation Facilities

Universal design needs to accommodate and satisfy all different groups at the same time. It is suggested that communal areas shall be provided to accommodate family activities. Compared to children and the elderly, youth need less special care but residential development planning must consider their needs, especially for social and recreational spaces. The following factors should be considered:

- It is preferable that social facilities, such as youth centres, cafes, restaurants, karaokes, gyms, game centres and sports facilities, shall be available within walking distance of residential areas.
- It is preferable that neighbouring open space and social space shall be available for residential developments, to help create a sense of identity and integrate different groups.

2.1.2 青少年

青少年文娱设施

通用设计需要照顾及满足所有不同的居民组别，建议应为家庭活动提供公共设施。若将青少年与儿童和长者相比，前者较少需要特别照料。不过，住宅发展规划必须考虑他们的需要，尤其是社交及文娱空间。规划过程中应考虑以下措施:

- 社交设施例如青少年中心、小餐厅、餐馆、卡拉OK、健身室、游戏中心及运动设施最好靠近住处，在步行可达的范围内。
- 在住宅区附近，最好设有露天场所及社交空间，以帮助建立邻里亲情，融合不同的居民组别。

2

2.1

2.1.3 长者与残障人士

长者与残障人士的畅达设计

长者和残障人士需要特殊照顾和特别的畅达设计。建议实行以下的规划方案:

- 建议为社区设施，包括老人中心、街市、商店、餐馆及为长者和残障人士而设的医护中心，设置无障碍的通道。
- 建议为残障人士提供易达的交通系统，基本设计原则如下:

(1) 建议公交车站应设于靠近住宅区的入口。

(2) 建议公交车站的指示牌用对比色、字体清晰且足够大。指示牌文字应设有凸起的盲文及字母，其高度应离地面850～1200mm。指示牌表面的照明光度应不少于120lx。

(3) 若有垂直交通系统，建议结合设置无障碍升降机。

(4) 建议在沿途提供座位，并在座位边设轮椅停放处。同时，设置雨棚或绿荫，供长者和残障人士歇息。

2.1.3 The Elderly and People with Disabilities

Accessibility for the Elderly and People with Disabilities

The elderly and people with disabilities need particular care and special accessible design features. To achieve accessibility design, the following planning solutions are recommended:

- It is recommended that barrier-free access shall be provided to community facilities including elderly centres, markets, shops, restaurants, and healthcare centres for the elderly and people with disabilities.
- It is recommended that accessibility to transport shall be provided for people with disabilities. The essential design approaches are as follows:

(1) It is recommended that bus stops shall be located close to the entrances of residential developments.

(2) It is recommended that bus stops signboards shall be in contrasting colours with legible text of sufficient size, Braille and raised letters at a height of between 850mm and 1200mm above the ground level. Illumination of no less than 120 lx shall be provided on the face of signboards.

(3) Where vertical circulation is needed, it is recommended that accessible lifts shall be provided.

(4) It is recommended that seating with wheelchair parking spaces, and overhead shading shall be provided along paths to enable the elderly and people with disabilities to rest.

Accessible lifts for vertical circulation
上落楼层的方便易达升降机

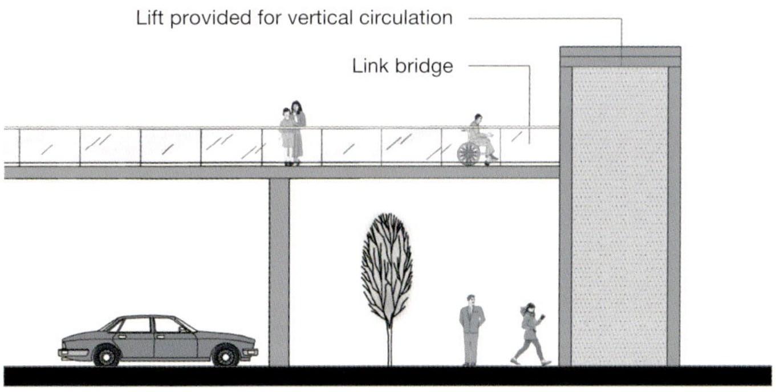

2.1

(5) It is recommended that accessible paths shall be provided to connect residential developments and bus stops. Preferably, shading shall be provided for accessible paths as part of an all-weather pedestrian system.

(6) It is recommended that audible signals shall be provided to indicate the route directions of approaching buses.

(7) It is recommended that bus stops shall have shelters and seating for the elderly and people with disabilities.

(8) It is recommended that at least one waiting space of more than 1200mm x 800mm shall be provided for people in wheelchairs.

(9) It is recommended that drop-off bays shall be located close to the entrances of residential developments. It is recommended that the depth of a drop-off bay shall be at least 3600mm wide, including a 1200mm wide access zone with dropped kerbs for people in wheelchairs.

(10) In high traffic areas and where circumstances allow, it is recommended that vehicular transport and pedestrian system shall be separated.

(11) Design of external areas and paths is discussed in chapter 3.2.

A bus stop with a shelter
有上盖的公交车站

Covered walkway
有盖的人行通道

A bus stop with a shelter and wheelchair parking space
有盖公交车停靠站并设有轮椅停放处

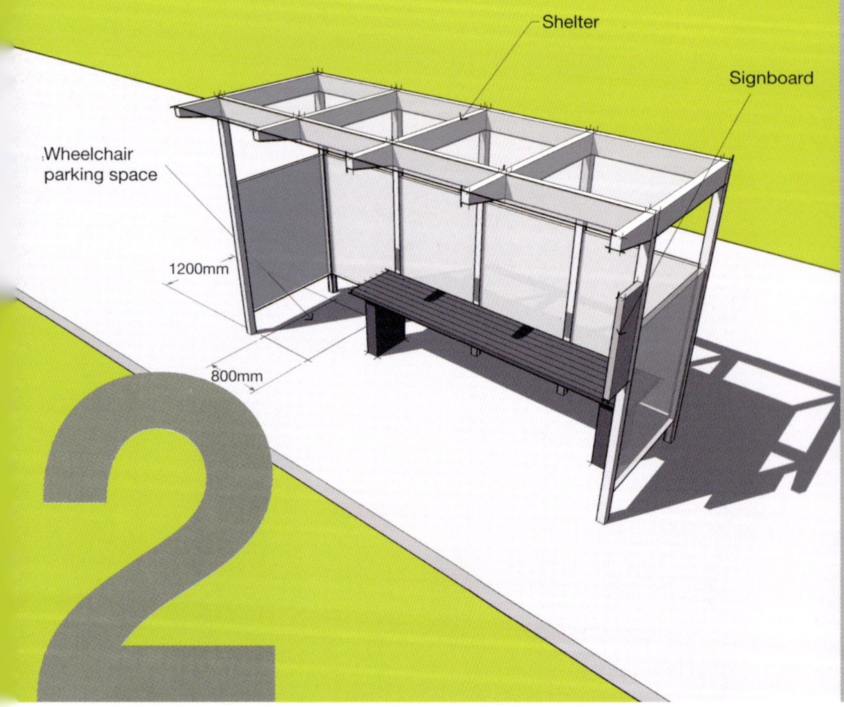

(5) 建议提供无障碍通道以连接住宅区及公交车站，并最好设有雨棚，使通道成为全天候行人系统的一部分。

(6) 建议提供声音信号提示到站巴士前往的方向。

(7) 建议在公交车站设置雨棚和座位，供长者和残障人士使用。

(8) 建议设置至少有一个大于1200mmx800mm的轮椅停靠空间，供轮椅人士停靠等候公交车。

(9) 建议汽车落客处应靠近住处入口。其深度至少应有3600mm宽，包括1200mm宽的轮椅通道与下斜路缘石相连。

(10) 在交通繁忙地区和情况许可下，建议汽车道路系统与人行道路系统分隔。

(11) 室外场地及通道的设计详见第3.2节。

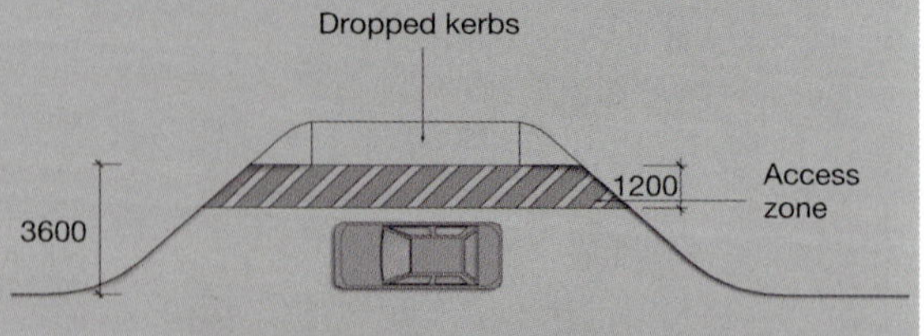

An access zone for a drop-off bay
落客处的使用范围

2.1

长者与残障人士的卫生健康

下列设计要点与长者和残障人士密切相关:

- 医护设备最好靠近住宅区。
- 住宅区附近最好设置护理机构。
- 最好设置户外和室内的健体设施。[参阅第5章]

长者与残障人士的社交设施

社交生活对包括长者在内的每一个人都很重要。由于大部分长者已退休，日常社交接触可能减少，不免孤单。户外的社交和交谊空间可以扩展年长居民的社交机会。

设计户外的社交和交谊空间时，建议应为长者和残障人士设置无障碍设施。第2.2.5节概述有关露天场所的规划设计建议措施。

Health and Care for the Elderly and People with Disabilities

The following factors have particular relevance for the elderly and people with disabilities:

- It is preferable that healthcare facilities shall be available near residential developments.
- It is preferable that nursing facilities shall be available near residential developments.
- It is preferable that indoor and outdoor health-promoting equipments shall be provided. [see chapter 5].

Social Considerations for the Elderly and People with Disabilities

Social life is important to everyone, including the elderly. As most elderly are retired, their access to casual social contact may be reduced, leaving them lonely. Outdoor social space can enhance social opportunities for elderly residents.

It is recommended that outdoor social space shall provide accessible settings for the elderly and people with disabilities. Details of open space planning will be discussed in paragraph 2.2.5.

Outdoor social space
户外社交地点

2.2 SPATIAL DESIGN IN HONG KONG CONDITIONS

2.2.1 Spatial Design of Bedrooms for Children in Wheelchairs

It is recommended that reach range shall be considered for children in wheelchairs. Forward or side reach ranges are recommended as follows:

- 500mm to 900mm for 3 to 4 years old children
- 450mm to 1000mm for 5 to 8 years old children
- 400mm to 1100mm for 9 to 12 years old children

Furniture Layout

- It is recommended that shelves and closets shall be considered in relation to children's reach range.
- Where window guards are not provided and children are present, it is preferable that furniture shall not be located near windows.
- It is recommended that a space of at least 900mm wide shall be provided on at least one side of a bed.
- It is recommended that a turning space of at least 1500mm x 1500mm should be provided in bedrooms. Where it is difficult to provide the turning space, a passageway of at least 900mm wide shall be provided as an access to all furniture and to the door.
- It is recommended that task lighting shall be provided at likely reading places.

2.2

2.2 符合香港情况的空间设计

2.2.1 使用轮椅的儿童卧室空间设计

建议可及范围应全面顾及坐轮椅的儿童，前方或侧面的可及范围如下：

- 3至4岁的儿童，可及范围500～900mm
- 5至8岁的儿童，可及范围450～1000mm
- 9至12岁的儿童，可及范围400～1100mm

家具摆设

- 建议架子和橱柜的位置应顾及儿童的可及范围。
- 如没有窗防护栅栏和有儿童居住，家具最好不要放置在窗旁。
- 建议在卧床的一边，至少留有900mm的空间。
- 建议卧室内至少应设有1500mmx1500mm的轮椅旋转空间。如不能提供此空间，则至少留有宽达900mm的通道，可通达所有家具及房门口。
- 建议在可能成为阅读地方的位置设置指向工作灯。

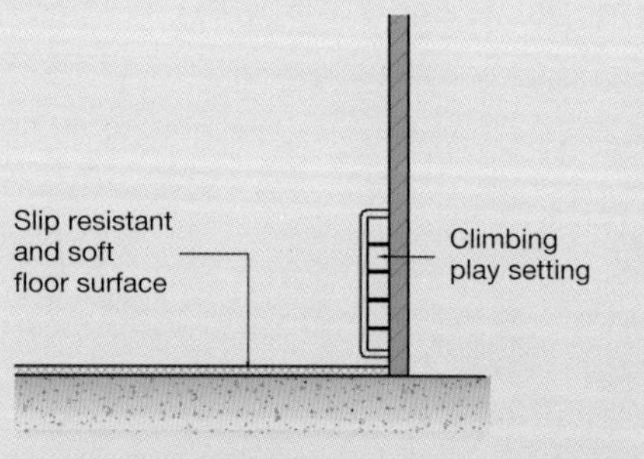

Floor surface for wall-mounted play equipment
有攀爬游玩设施时的地板设计

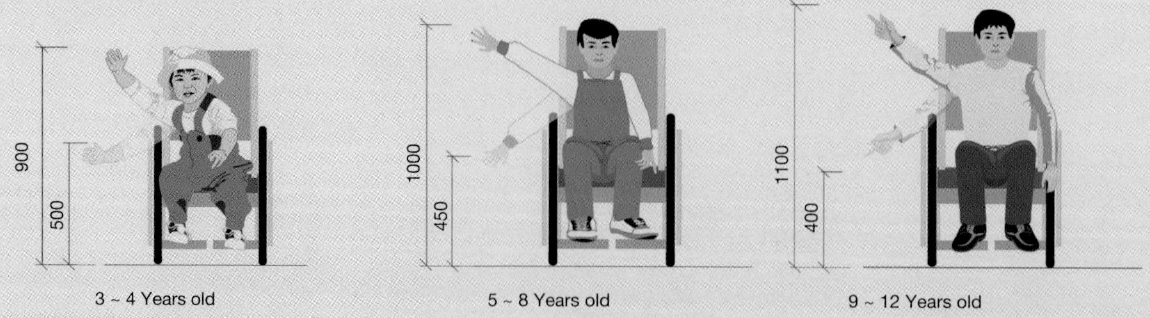

Children's reach range
儿童可及范围

推荐方案

- 在资源许可的情况下，最好在儿童卧室设置配合年龄的摆设和设有安全措施的游玩空间。
- 游玩设备如设在墙上，建议应检查装置是否稳妥及有足够承载力，地板也应防滑和具有消减冲击力的能力。建议安装在墙上的游玩设备不应靠近窗户，以及有尖角或凸出物。
- 家具应该耐用和稳固，高度和摆设位置应可调校。

Good Practice

- Where resources allow, it is preferable that children's bedrooms shall have age-appropriate layout and a hazard free play area.
- Where play equipment is wall-mounted, it is recommended that the wall shall be checked for stability and mounting strength and that floor finishes shall be slip-resistant and impact absorbing. It is recommended that wall-mounted play equipment shall not be located close to windows and furniture which has sharp corners or protrusions.
- It is preferable that furniture shall be durable, stable and adjustable in respect of height and layout.

2.2.2 长者和残障人士卧室的空间设计

长者和残障人士，尤其轮椅人士，对房间内的活动空间会有更高要求。

- 为确保有足够空间让轮椅通过，建议门的净宽度至少应为850mm。
- 建议在卧床的一边，至少留有900mm的净空间。
- 最好在室内提供一处1500mmx1500mm的净空间作为轮椅旋转空间。
- 建议在门开启的方向，在靠近门把手一侧的墙面，应留有500mm宽的墙面净空，作为轮椅使用者开启门的操作空间。

2.2.2 Spatial Design of Bedrooms for the Elderly and People with Disabilities

The elderly and people with disabilities, especially people in wheelchairs, have higher requirements of room size for mobility.

- It is recommended that for wheelchair access, the clear width of a doorway shall be at least 850mm.
- It is recommended that a clear space of at least 900mm wide shall be provided on at least one side of a bed.
- Preferably, a clear space of 1500mm x 1500mm shall be provided inside a bedroom, to facilitate a turning of a wheelchair.
- It is recommended that a manoeuvring clearance for front approach to doors shall not be less than 500mm wide.

Options for elderly bedroom size:
长者卧室的大小可有下列选择:

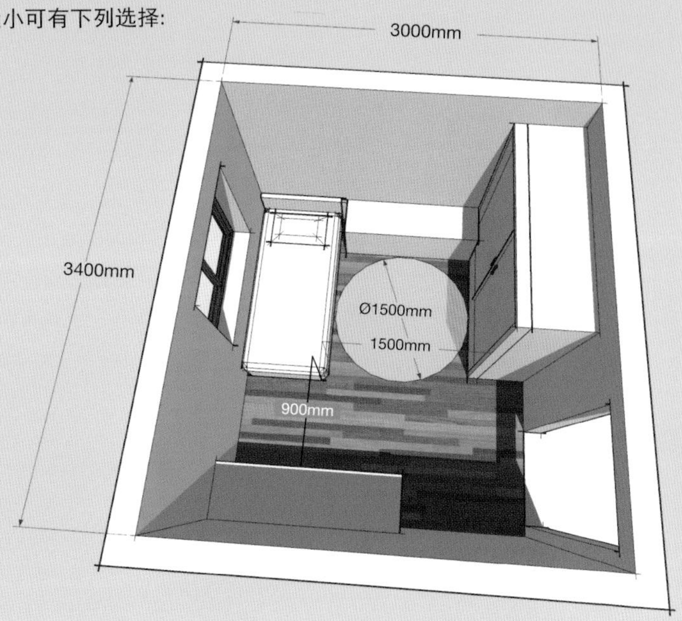

Elderly bedroom layout option 1
长者卧室方案 1

Options for elderly bedroom size:
长者卧室的大小可有下列选择:

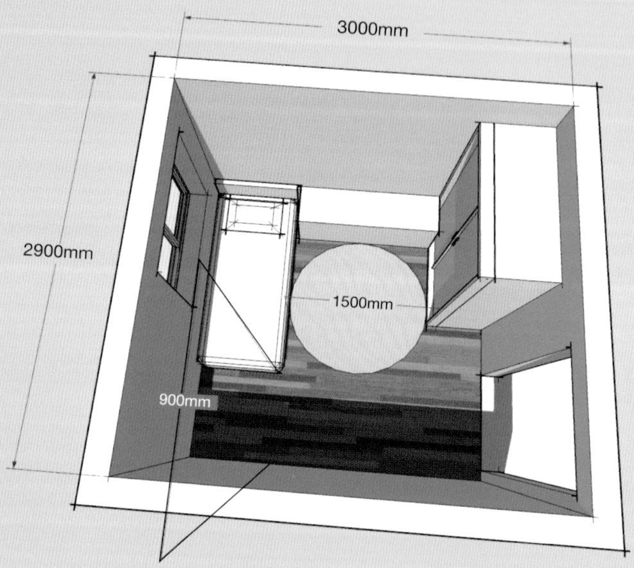

Elderly bedroom layout option 2
长者卧室方案 2

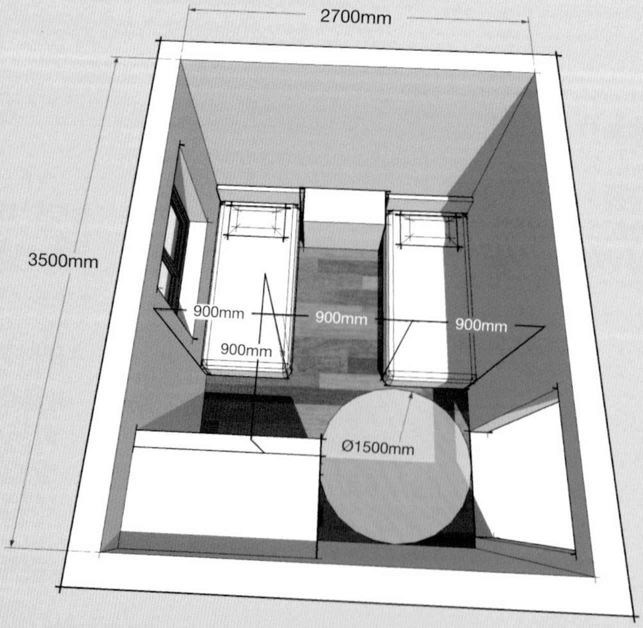

Elderly bedroom layout option 3
长者卧室方案 3

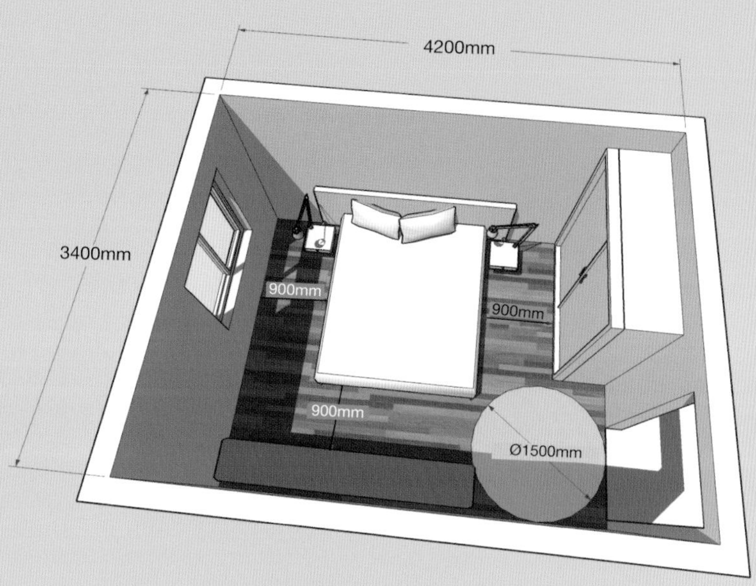

Elderly bedroom layout option 4
长者卧室方案 4

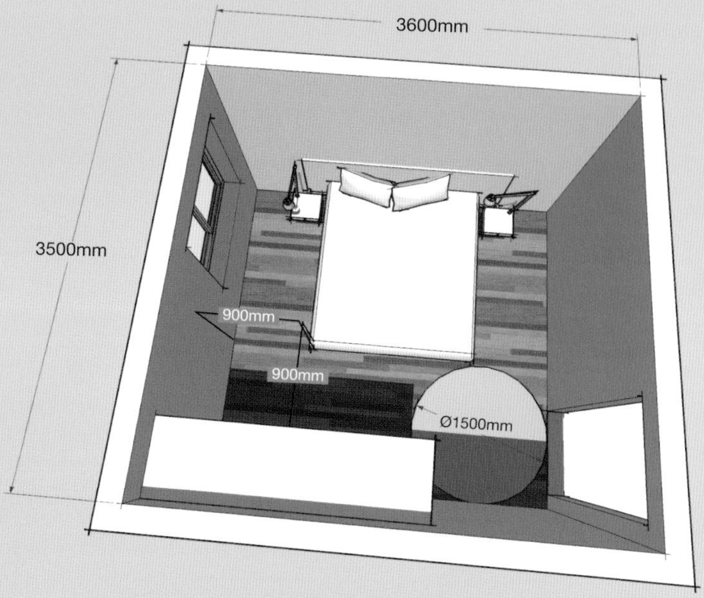

Elderly bedroom layout option 5
长者卧室方案 5

2.2.3 Spatial Design of Living / Dining Rooms for the Elderly and People with Disabilities

Furniture

- Where resources allow, it is recommended that furniture shall be provided which is sturdy, stable and without sharp or protruding edges and corners.

Furniture Layout

- For safety and ease of use by the elderly, people with disabilities or visual impairments, it is recommended that furniture layout shall be orderly, with adequate manoeuvring space.
- It is recommended that children's play space shall be provided in the Living Room area to foster interaction and socialization between children, caregivers and the elderly.

Good Practice

- Where circumstances allow, it is preferable that furniture shall not be located close to windows.
- It is preferable to provide a wheelchair parking space with a minimum area of 800mm x 1200mm beside a sofa, opposite to likely TV location.
- It is preferable that a manoeuvring space or turning circle of at least 1500mm x 1500mm shall be provided in front of a window.
- To facilitate reading or visually demanding tasks, it is recommended that task light shall be provided at likely wheelchair parking spaces.
- To facilitate access by the visually impaired, it is recommended that furniture shall be in contrasting colours, tones or materials.

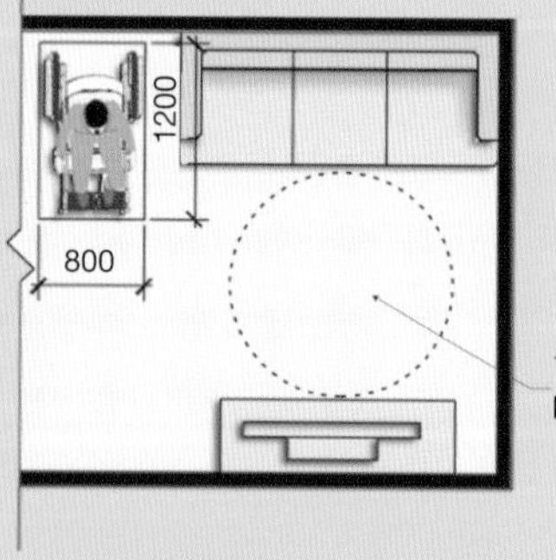

Wheelchair parking space and manoeuvring space in a Living/Dining room
客饭厅的轮椅停放处及操作空间

2.2

2.2.3 长者和残障人士客/饭厅的空间设计

家具

- 在资源许可的情形下，建议应设置结构稳固的家具，并且家具不应有尖角或凸出的边角。

家具摆设

- 为方便长者、残障或视障人士使用及安全起见，建议家具的摆设应有条理，使用时有足够的轮椅操作空间。
- 建议应在客厅提供儿童游玩空间，促进儿童、照顾者及长者之间的交流与活动。

推荐方案

- 在情况许可下，家具最好不要靠近窗户。
- 沙发旁或电视机对面最好设有轮椅停放处，面积至少为800mmx1200mm。
- 最好在窗户前至少留有1500mmx1500mm的轮椅操作空间或转动空间。
- 建议在轮椅停放处设置指向工作灯，方便阅读和进行依赖视力的工作。
- 建议家具颜色使用对比色，并可在物料上区分，方便视障人士使用。

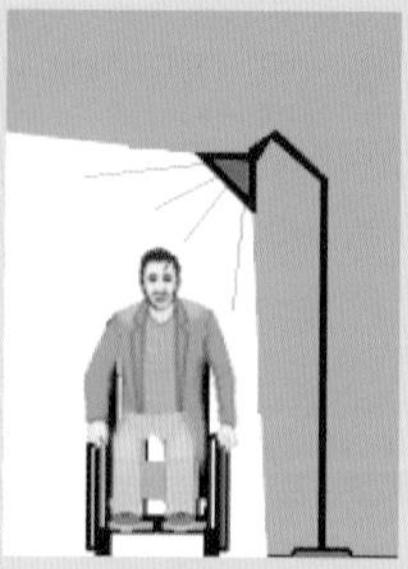

Task lighting for wheelchair parking space
轮椅停放处工作灯

2

2.2

花槽

接触自然界是居民生活的一部分，在家中摆放植物，可为经常在家的人增添生活姿彩。

- 为方便轮椅人士活动，花槽的高度建议应介乎400 ~ 1000mm之间。
- 避免种植有刺和有毒的植物，以免受伤。

Planters

Contact with nature is an important part of the living experience. Provision for plants at home can enrich the lives of people who may spend most of their time there.

- To facilitate activity by people in wheelchairs, it is recommended that planters shall be at a height of between 400mm and 1000mm.
- To avoid injury, it is preferable to exclude plants with thorns, spikes or poisonous qualities.

2.2.4 多代同堂住宅平面设计

香港家庭传统上会长幼同住。为了避免冲突、容纳自主和私隐，最好能够局部分隔长幼两代的生活空间，并通过共用地方将两代居住空间相连。此类住宅可能采用的平面规划如下：

2.2.4 Flat Layouts for Families

It is traditional in Hong Kong for younger and older generations to live together. To minimize conflict and facilitate independence and privacy, it may be desirable to partially separate older and younger generations, using common areas to connect them. Possible flat layouts are shown:

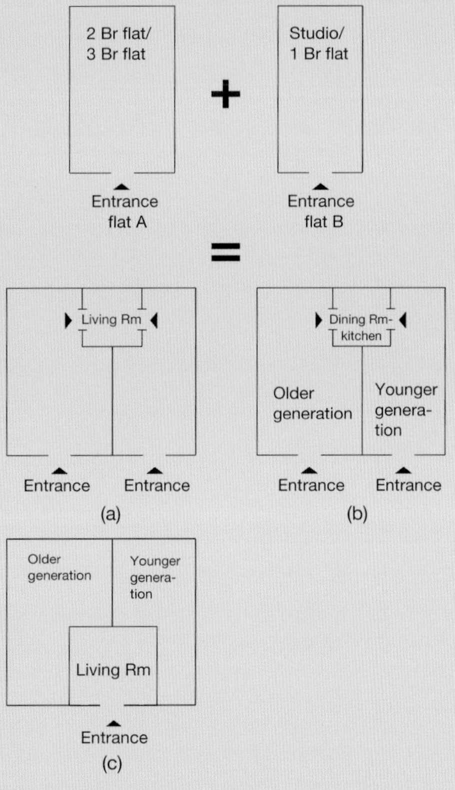

(a) Separate entrances, living room as common area
(b) Separate entrances, kitchen/dining rooms as common area
(c) One entrance, living room as common area

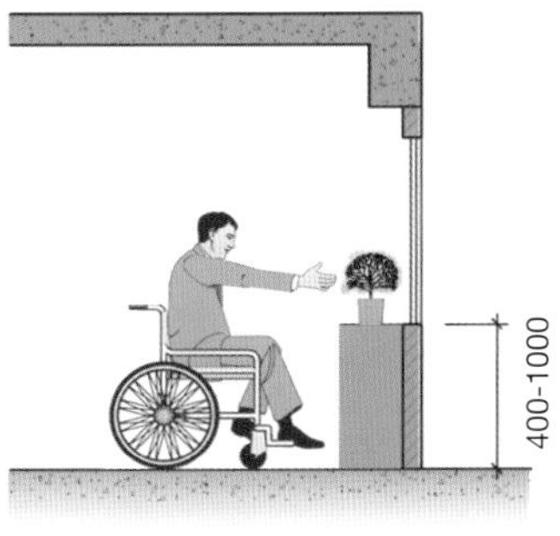

Location of planters in living rooms
客厅花槽位置

Good Practice

- Where circumstances allow, it is preferable that furniture, such as tables, closets, shelves, and bathroom fixtures shall be adjustable in respect of height, so it can be universally usable.
- Where circumstances allow, it is recommended to provide movable partitions to a Master Bedroom so that it can be changed from a single-bed to double-bed room to allow adaptability of layout.

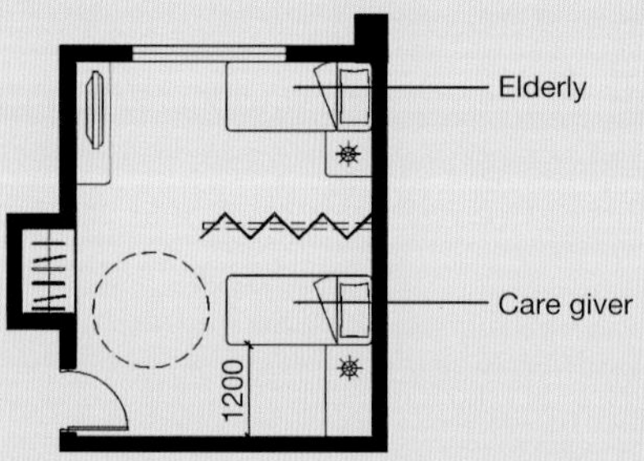

A movable partition for a master bedroom
主人房内可迁移的间隔

推荐方案

- 在情况许可下，餐桌、橱柜、架子及沐浴设备等家具最好应可调校高度，以方便不同人士使用。
- 在情况许可下，建议主人房应设置可移动的间隔，以便使单人房可改为双人房，灵活使用。

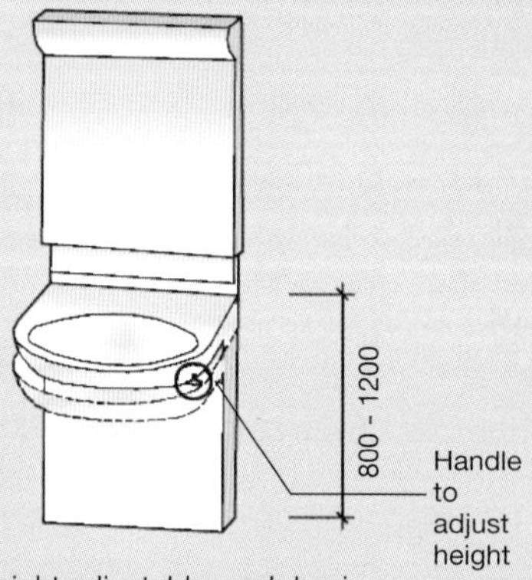

Height adjustable wash basin
可调校高度的洗涤盆

2.2

2.2.5 Spatial Design for Neighbouring Playgrounds and Open Space

Accessibility and safety are essential issues to neighbouring playgrounds and open space. It is preferable that playground standards shall create equal opportunities for children with disabilities, and open space shall provide accessible social space.

2.2.5.1 Neighbouring Playgrounds

Entrances to Neighbouring Playgrounds

- It is recommended that entrances shall be located away from traffic and accessible to children with disabilities.
- It is recommended that the clear width of an entrance shall be not less than 900mm, ideally 1500mm.
- It is recommended that entrances shall be connected to accessible pathways, connecting all play components, accessible elements and spaces.
- It is recommended that manoeuvring space of at least 1500mm x 1500mm shall be provided on both sides of an entrance.

2.2.5 邻近游乐场及露天场所的空间设计

邻近的游乐场和露天场所必须方便出入、容易到达，并且安全。游乐场的规格最好能为残障儿童提供平等机会，而露天场所则应提供方便易达的社交空间。

2.2.5.1 邻近游乐场

通往邻近游乐场的入口

- 建议入口应远离交通要道，并且方便有残障的儿童到达。
- 建议入口的净宽度至少要有900mm，1500mm则为最理想。
- 建议入口要有通道接驳，并连接所有游玩设施、接驳元素和空间。
- 建议入口两旁均应设有至少1500mmx1500mm的轮椅操作空间。

2

Guidelines 规划和空间设计指南

2.2

推荐方案

- 在情况许可下，建议游乐场应有护卫员巡逻。
- 建议在儿童游乐场的入口，标示使用者年龄和安全使用警告的指示牌。

邻近游乐场的通道和斜道

- 通道和斜道的设计详见于第3.2节和第3.4节。
- 建议无障碍通道应连接所有游玩设施。
- 游乐场通道最少净宽度应有1500mm。如果难以提供1500mm宽的通道，宽度至少要有900mm，并于每10m在通道旁设有至少1500mmx1500mm的轮椅操作空间。
- 建议无障碍通道表面必须防滑。游玩设施的底部和周边通道的表面应有减缓冲击力的功能，并应定期维修。

邻近游乐场游玩设施的地面

- 建议地面撞击范围应有减缓冲击力的功能。
- 建议游玩设施底部及周边应设有安全防护地面。
- 减缓冲击力的地面应符合下列安全标准:

(1) EN1176-Part1~7[BSI-EN(2000),Europe]
(2) ASTM F1487-01[ASTM (2001), U.S.]

邻近游乐场的游玩设施

- 建议邻近游乐场应顾及残障儿童的需要，设有多元化的游玩设施。
- 建议在地面为视障儿童提供无障碍游玩设施，例如触板和可发声的设施。
- 建议高架的游玩设施应顾及畅达设计措施。
- 建议游玩区应按年龄组别分开，所有设施均应显示适合的年龄组别。

Chimes panel

Reach panel

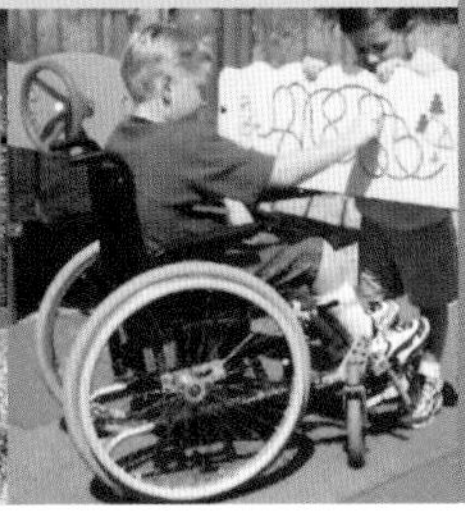

Good practice

- Where resources allow, it is recommended that security patrols shall be provided at playgrounds.
- It is recommended that user age range and precaution signage for play safety shall be provided at the entrance of a children playground.

Paths and Ramps in Neighbouring Playgrounds

- Design of accessible paths and ramps are discussed in Chapter 3.2 and 3.4
- It is recommended that accessible paths shall connect all play components.
- It is recommended that the clear width of a path in a playground shall be at least 1500mm. Where it is difficult to provide a path of 1500mm in width, the minimum width of the path should be 900mm and a manoeuvring space of at least 1500mm x 1500mm should be provided every 10m along the path.
- It is recommended that the surface of accessible paths shall be slip-resistant. The surface of the paths under and around playground equipment should have impact absorbing ability. Regular maintenance shall be applied.

Play Surfaces in Neighbouring Playgrounds

- It is recommended that the surface of impact areas shall have impact absorbing ability.
- It is recommended that safety surfacing shall be provided under and around play equipment.
- The performance of impact absorbing surfaces is considered in the following Safety Standards:

(1) EN1176 - Part 1~7 [BSI-EN(2000), Europe]
(2) ASTM F1487-01 [ASTM (2001), U.S.]

Play Components in Neighbouring Playgrounds

- It is recommended that diverse play components shall be provided in neighbouring playgrounds, with considerations of children with disabilities.
- To provide for children with visual impairments, it is recommended that accessible play components, such as touch panels and equipment with sounds, shall be provided at ground level.
- It is recommended that accessibility shall be considered for elevated play components.
- It is recommended that play area shall be separated according to age groups, with signs to indicate to which age groups components are suited.

2.2

Good Practice

- It is recommended that at least one of each type of play components, which is provided at ground level in the play area, shall be on an accessible pathway.
- It is recommended that at least 50% of the elevated play components shall be on an accessible pathway.

Transfer Steps and Ramps in Neighbouring Playgrounds

- To enable children who use wheelchairs to leave the wheelchair or other mobility device and transfer or be lifted onto elevated play components, it is recommended that ramps, transfer platforms or transfer steps shall be provided.
- Where there is a change in level, ramps or transfer steps shall be provided on accessible paths.
- It is recommended that a transfer platform shall be 350 - 450mm in height, not less than 610mm in width, and not less than 350mm in depth.
- It is recommended that a transfer step shall be not more than 200mm in height, not less than 610mm in width, and not less than 350mm in depth.
- To allow for wheelchair parking, it is recommended that a space of not less than 1200mm x 800mm shall be provided beside the first transfer platform.
- It is recommended that handrails shall be provided on both sides of ramps and elevated corridors.
- It is recommended that handrails shall be between 25mm and 40mm in diameter or width.

推荐方案

- 建议每类设于游玩场地的游玩设施中至少有一个设施位于无障碍通道上。
- 建议至少要有一半的高架游玩设施设于无障碍通道上。

邻近游乐场的转移梯级及斜道

- 建议设置斜道、转移平台或转移梯级，供使用轮椅或其他行动辅助器的儿童摆放这些工具后，转移或升高至高架的游玩设施。
- 如果高度有变，斜道及转移梯级应设于无障碍通道上。
- 建议转移平台的高度应在350 ~ 450mm之间，宽度应不少于610mm，深度应不少于350mm。
- 建议转移梯级的高度不应超越200mm，宽度不少于610mm、深度不少于350mm。
- 建议在第一个转移平台侧，应设有1200mmx800mm的空间，供轮椅停放。
- 建议在斜道及高架走廊两旁设置扶手。
- 建议扶手的直径或宽度应介于25 ~ 40mm之间。

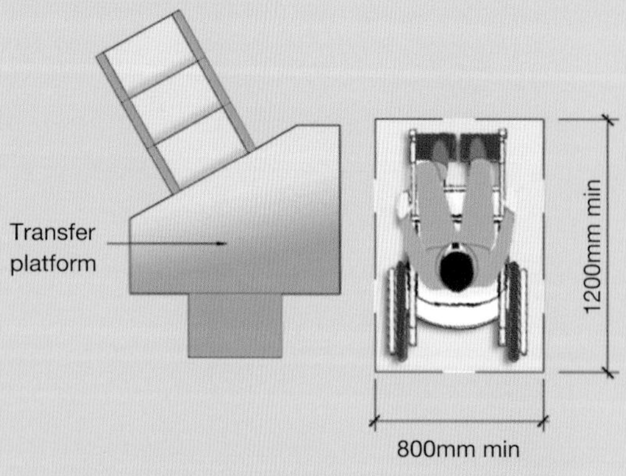

Wheelchair parking space beside a transfer platform
设于转移平台旁的轮椅停放处

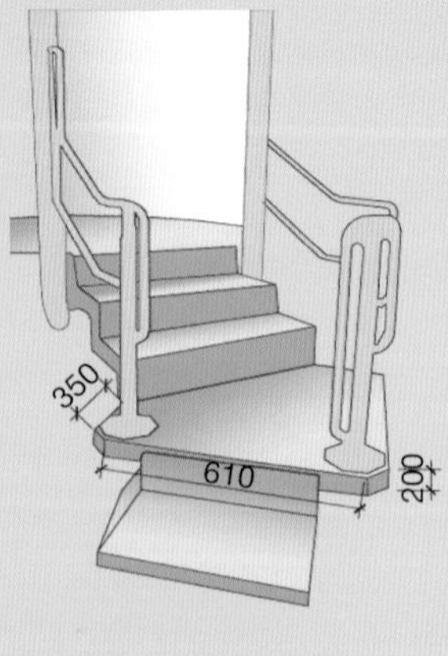

A transfer step for a play component
游乐设施的转移平台

2

2.2

推荐方案

- 如需要使用转移的方法通往每一个高度，建议在转移平台和转移梯级处提供转移承托。
- 如果使用转移系统，建议使用爬行管道和隧道等创新的通道，连接各种高架游玩设施。

邻近游乐场的游玩桌

- 建议预留予轮椅的膝盖净空间不应少于650mm，宽度不少于800mm，深度不少于450mm。
- 建议游玩桌不应高于800mm。
- 建议游玩桌边沿应圆滑。
- 建议游玩桌应设于荫棚下。

推荐方案

- 建议在游玩桌边应设有一个细小凹进处，用以摆放手杖、雨伞或袋子。
- 建议游玩桌边旁应设有挡条，以防游玩物品堕下。

邻近游乐场的照明

建议游乐场应有足够光线照明。照明度最好在75～120lx。邻近地区的照明越大，则应设置越高的照明度。

Good Practice

- At each level where transfer is the intended method of access, it is recommended that transfer support shall be provided on transfer platforms and transfer steps.
- When a transfer system is used, it is recommended that an elevated play component shall be connected to other elevated play components through innovative accessible routes such as crawl tubes and tunnels.

Play Tables in Neighbouring Playgrounds

- It is recommended to provide wheelchair knee clearance with a height not less than 650mm, a width not less than 800mm and a depth not less than 450mm.
- It is recommended that the top of a play table shall not be higher than 800mm.
- It is recommended that the edge of a play table shall be rounded.
- It is recommended that a play table shall be shaded.

Good Practice

- To facilitate the placement of walking canes, umbrellas or bags, it is recommended that a small recess shall be provided at the edge of a play table.
- To prevent play materials from falling, it is recommended that an upstand shall be provided at the edge of a play table.

Lighting of Neighbouring Playgrounds

It is recommended that lighting shall be provided at a sufficient illumination level in playground. Preferably, the illumination level shall be between 75 and 120 lx. The greater the lighting of the adjacent area is, the higher the illumination level shall be chosen.

A small recess for a play table
游玩桌的细小凹进处

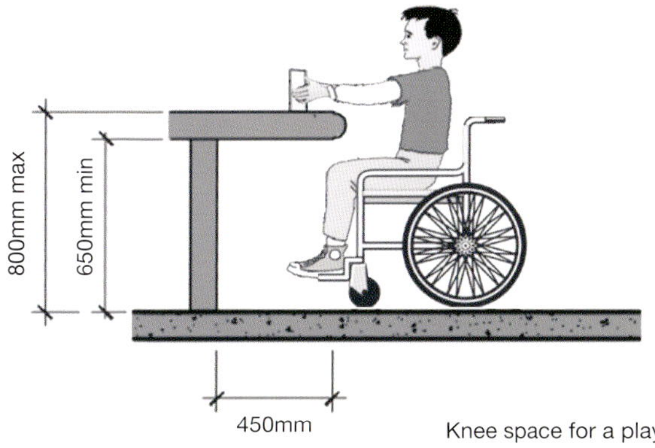

Knee space for a play table
游玩桌的膝部空间

2.2

2.2.5.2 Open Spaces

Neighbouring open space can act as an extension of living space. It can provide social space for both the young and the elderly. To create an open space capable of satisfying a wide range of potential users, it is preferable to give consideration to people with disabilities at all ages.

Entrances to Open Space

- It is recommended that entrances shall be clearly indicated.
- It is recommended that the clear width of entrances shall be not less than 1200mm, ideally more than 1500mm.
- It is recommended that signboards shall be provided to exit routes.
- To facilitate use by the visually impaired, it is recommended that signboards at entrances shall be in contrasting colours. Braille and raised letter / characters shall also be provided on the signboards.

Good Practice

- Where circumstances allow, it may be desirable that open space shall be located in the centre of residential areas.
- To help people with visual impairments to find their ways and identify locations, it is desirable to provide scented plants outside the entrances.

Paths for Open Spaces

- Design of accessible paths is discussed in Chapter 3.2.
- It is recommended that accessible paths shall connect main entrances to all activity places.
- To help people with visual impairments, it is recommended that distinctive paving textures shall be provided to different activity zones.
- To help people with visual impairments, it is recommended that tactile positional tiles shall be provided in front of signboards.

2.2.5.2 露天场所

如要扩大居住的空间，可从邻近的露天空间着手，这些空间可为长幼提供社交场所。如要露天场所有效地切合各种使用者的需要，则须考虑不同年龄残障人士的需要。

露天场所入口

- 建议在入口设有清晰的方位指示。
- 建议入口的净宽度不少于1200mm，最理想值大于1500mm。
- 建议使用指示牌显示出口方向。
- 为方便视障人士，建议指示牌的颜色选用对比色，并设有凸起的盲文及字母/文字。

推荐方案

- 在情况许可下，露天场所最好设置于住宅区中心地带。
- 入口外面最好种植有香气的植物，协助视障人士寻找和辨认地点。

露天场所通道

- 无障碍通道的设计详见于第3.2节。
- 建议用通道连接主要入口及所有活动地点。
- 为协助视障人士，建议在不同的活动点铺设不同的地面物料。
- 为协助视障人士，建议在指示牌前设置方位引路砖。

推荐方案

- 为方便视障人士辨认方向，各类通道旁最好应种植不同气味的植物。

露天场所的社交区

- 建议在露天场所设置社交区。
- 建议在社交区设置雨棚和上盖。
- 建议在社交区设置长椅，给成人的座位，应设有扶手及椅背，以增加舒适度。
- 建议在社交区的长椅子旁设置至少800mm x1200mm的轮椅停放处。

推荐方案

- 建议在社交区的周边栽种植物，增添视觉趣味。
- 为安全理由，建议社交区不应被植物遮蔽。
- 在夜间开放的社交区，建议应提供足够的照明，最好在75～120lx之间。

露天场所的植物

- 为安全起见，尤其是为了视障人士，建议树木如有低生横枝便应修剪，或将之种植于远离无障碍通道的地方。
- 建议应考虑利用不同植物，在四季营造一个全年充满视觉趣味的景致。

推荐方案

- 花卉植物建议要选用对比色，以供视障人士欣赏。
- 建议避免种植有害和有毒的植物。

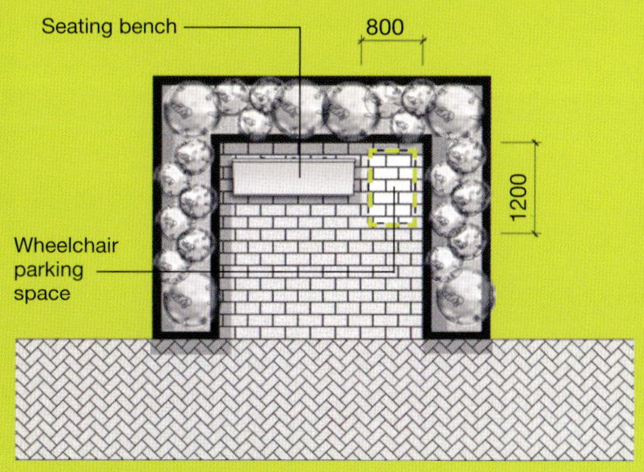

Wheelchair parking space at a social corner
社交角的轮椅停放处

Good Practice

- To facilitate people with visual impairments to identify directions, it is desirable that flowery plants with different fragrances shall be planted on the sides of paths.

Social Corners in Open Space

- It is recommended that social corners shall be provided where there is open space.
- It is recommended that shadings and covers shall be provided at social corners.
- It is recommended that benches shall be provided in social corners. Seats for adults shall have armrests and seatbacks to increase comfort.
- In social corners, it is recommended that a wheelchair parking space of at least 800mm x 1200mm shall be provided beside each bench.

Good Practice

- For visual interest at social corners, it is recommended that planting shall be provided.
- For security, it is recommended that planting shall not conceal a social corners from their surroundings.
- If social corners are accessible at night, it is recommended that illumination shall be provided at a sufficient level, preferably between 75 and 120 lx.

Plants in Open Space

- For safety, particularly of the visual impaired, it is recommended that trees with low branches shall be pruned or planted away from accessible paths.
- It is recommended to consider the seasonal appearance of planting in order to create a visually interesting landscape all year round.

Good practice

- For the pleasure of the visual impaired, it is recommended to provide plants and flowers in contrasting colours or tones.
- It is recommended that injurious and poisonous plants shall not be planted.

Cha 章

pter

8

Barrier-Free Access
无障碍设计

Barrier - Free Access

3.1 INTRODUCTION

As shown in the Universal Design Pyramid, the elderly and people with disabilities have higher requirements for accessibility. This group of people are increasing in Hong Kong. In Hong Kong, the population is aging due to declining fertility and longer life expectancy. According to the report of the Census and Statistics Department [C&S], the elderly dependency ratio has increased from 124 in 1991 to 154 in 2001. C&S also projected that the proportion of those aged 65 and above will rise from 12% in 2003 to 27% in 2033. Furthermore, the disabled population represents 5% of the total population of Hong Kong, according to C&S' study in 2000. It includes 103,500 persons with physical handicap, 73,900 persons with visual impairments, 69,700 persons with hearing impairments and 18,500 persons with speech impairments. Therefore, more attention shall be paid to increase accessibility for the elderly and people with disabilities throughout the built environment.

The Hong Kong Government has provided statutory documents to guide barrier-free design. They include the Disability Discrimination Ordinance, implemented since 1996, the Residential Care Homes (Elderly Persons) Ordinance and the Design Manual - Barrier Free Access 1997. Based on these statutory documents, this chapter will introduce good practice for residential developments.

3.1 引言

正如通用设计金字塔显示，长者和残障人士对畅达设计有更高的要求。在香港，这一组别的人士正不断增加，香港的人口由于出生率下降和寿命延长而持续老龄化。根据政府统计处的报告，长者的受赡养比率从1991年的124增至2001年的154。统计处也预计65岁和以上的人口比例将从2003年的12%增至2033年的27%。此外，统计处于2000年的统计研究指出，残障人口占香港总人口的5%，包括103,500位身体残障人士、73,900位视障人士、69,700位听觉障碍人士及18,500语言障碍人士。因此为了长者和残障人士，我们应更加注重提高建筑环境的畅达性。

香港特区政府已为无障碍设计提供法定的指引文件，包括自1996年起实施的《残疾歧视条例》、《安老院条例》和《设计手册：畅通无阻的通道1997》。本章将以这些法定文件为基础，介绍住宅建设中的无障碍设计。

3.2 EXTERNAL AREAS AND PATHS FOR BARRIER-FREE ACCESS

3.2.1 Dropped Kerbs in External Areas

- It is recommended that dropped kerbs shall be provided at pedestrian crossings, at each end of footpaths and at building entrances, where appropriate.

3.2 无障碍室外场所和通道

3.2.1 室外下斜路缘石

- 建议在行人过路处、行人路末端及楼宇入口的适当地方设置下斜路缘石。

3.2

- It is recommended that dropped kerbs and kerb ramps shall be ramped at a gradient not steeper than 1 : 8, ideally 1 : 12.
- It is recommended that dropped kerbs shall be provided with a minimum length of 1200mm and a minimum width of 1300mm.
- It is recommended that a clear space of 1500mm x 1500mm for landing shall be provided behind the dropped kerbs.
- It is recommended that the surface of dropped kerbs shall be slip-resistant and be installed with tactile warning strips.

- 对于下斜路缘石及连接下斜路缘石的路缘坡度，建议斜度不应深过1:8，最理想斜度为1:12。
- 建议下斜路缘石的长度至少应有1200mm，宽度至少应有1300mm。
- 建议在下斜路缘石后，提供1500mmx1500mm的净空间，方便上落。
- 建议下斜路缘石表面应防滑，并装设有警告条纹的引路砖。

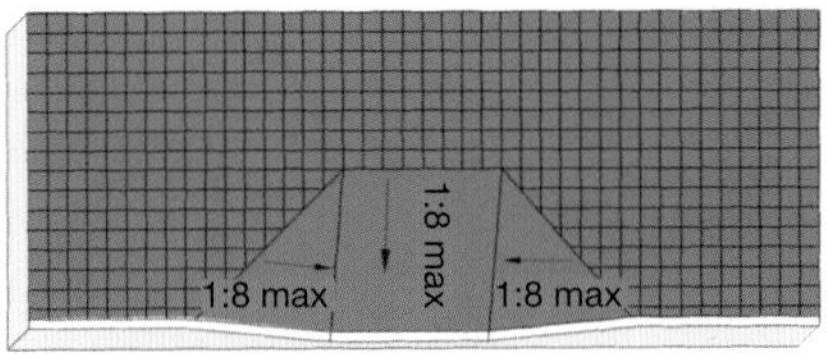

A dropped kerb for a path
通道下斜路线

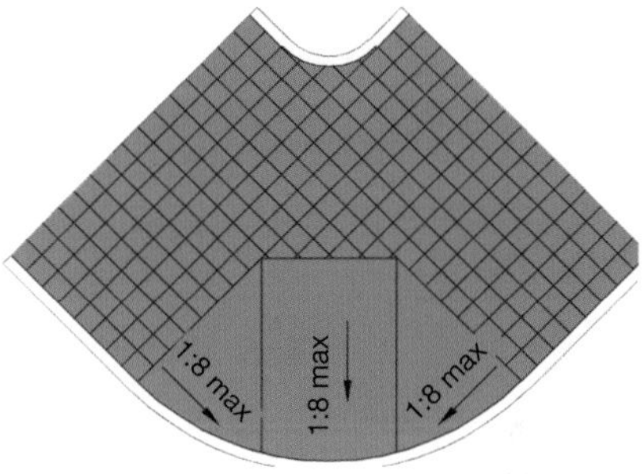

A dropped kerb at the corner of pavement
行人路角位的下斜路线

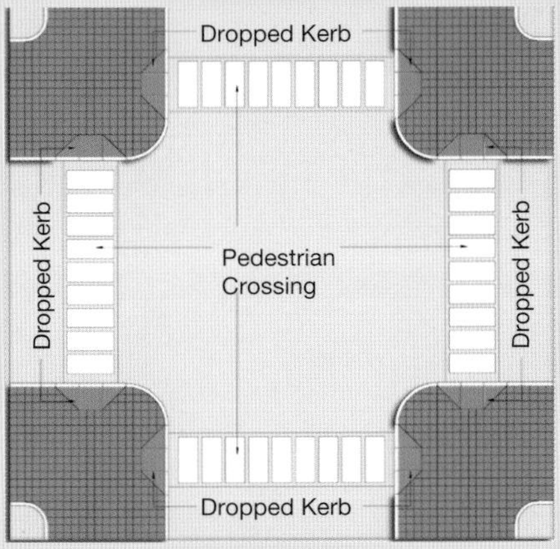

Dropped kerb for pedestrian crossings
行人路的下斜路缘

A dropped kerb with hazard warning tactile surface
有凹凸警告条纹的下斜路缘

3.2

3.2.2 Paths in External Areas

- In order to facilitate access by people in wheelchairs, it is recommended that the width of accessible paths shall be not less than 900mm and ideally 1500mm. If a path is less than 1500mm in width, it is preferable that a space of at least 1500mm x 1500mm shall be provided every 10m to accommodate the passing and the turning of wheelchairs.
- It is recommended that a tactile guide path shall be provided to one side of an accessible path.
- It is recommended that angles at intersections of accessible paths shall be splayed or rounded.
- In order to assist people with visually impairments, it is recommended that corners and edges of paths shall be clearly marked by using colour, brightness, textures or upstand edge.
- It is recommended that the surface of accessible paths shall be slip-resistant. Regular maintenance shall be provided.
- It is recommended that plants or planters beside an accessible path shall be located between 450mm and 600mm away from a tactile guide path.
- Paths should be as levelled as possible. It is recommended that the slope for an accessible path shall not exceed 1:12, and ideally not steeper than 1:25, especially for longer distances.
- It is recommended that lighting and seating shall be provided along accessible paths.

3.2.2 室外通道

- 为方便轮椅人士，建议通道的净宽不应少于900mm，最理想宽度为1500mm。如宽度少于1500mm，最好在每10m处提供面积至少1500mmx1500mm的空间以容纳轮椅通过和回转。
- 建议在无障碍通道的一边提供凹凸纹引路径。
- 建议无障碍通道的交叉口处宽度予以放大或者处理成圆角。
- 为协助视障人士，建议通道边缘位应以颜色、光度、质感和竖柱边显示。
- 建议无障碍通道的表面应该防滑，并应定期维修保养。
- 建议无障碍通道旁的植物或花槽，应离开引路径450~600mm。
- 通道应尽量平坦，建议无障碍通道的斜度不应超越1:12，最理想斜度应少于1:25，尤其是较长的通道。
- 建议在无障碍通道沿途设置街灯和座椅。

Manoeuvring space along path
通道旁的轮椅操作空间

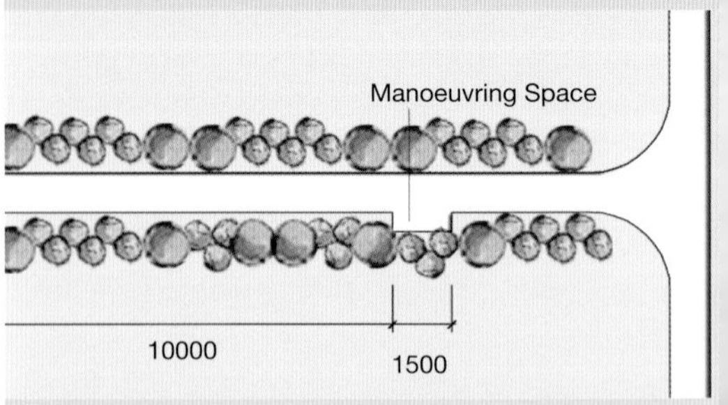

Guide path on one side of an accessible path
在方便易达通道旁的指引通道

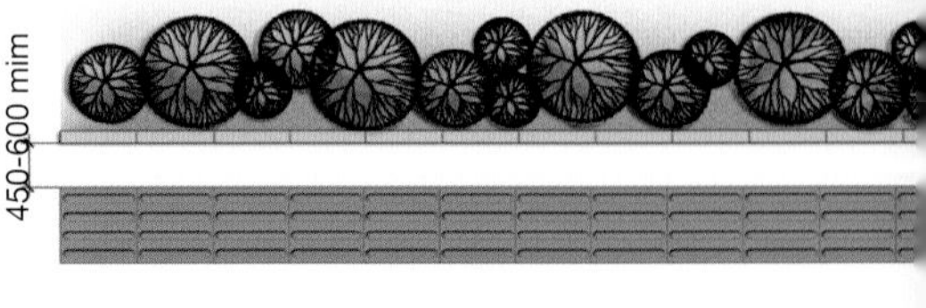

3.2

- It is recommended that an overhead clearance of 2m shall be maintained under tree limbs, signs and lights, over full width of accessible paths.
- Where drainage channel intersects accessible paths, it is recommended that gratings shall be provided. Channel grating slots shall not be parallel to the direction of major traffic. Openings in gratings shall be small enough to avoid trapping of crutches or wheels.
- In order to assist people in wheelchairs, it is recommended that loose gravel and sand shall not be used as the surface to accessible paths.
- Ideally, tactile warning tiles or surfaces shall be provided on the pavements around signposts, streetlights, trees, mailboxes, and other street furniture.
- It is recommended that shading / cover shall be provided to accessible patdl hs and seating areas.

- 建议沿无障碍通道和其沿途座位设置绿荫或雨棚。
- 建议沿无障碍通道的树枝、指示牌和路灯下的净空高度应有2m。
- 排水渠与无障碍通道交界的地方应设置格栅。建议渠道格栅的狭槽不应与主要交通方向平行，格栅的开口应尽量细小，以免拐仗或轮椅的轮子陷入。
- 为协助轮椅人士，建议无障碍通道的表面不应铺盖松散的砾石和砂粒。
- 在行人路上有指示牌、街灯、树木、邮箱和其他街道设施的位置，最理想应铺设有凹凸纹的危险警告引路砖。

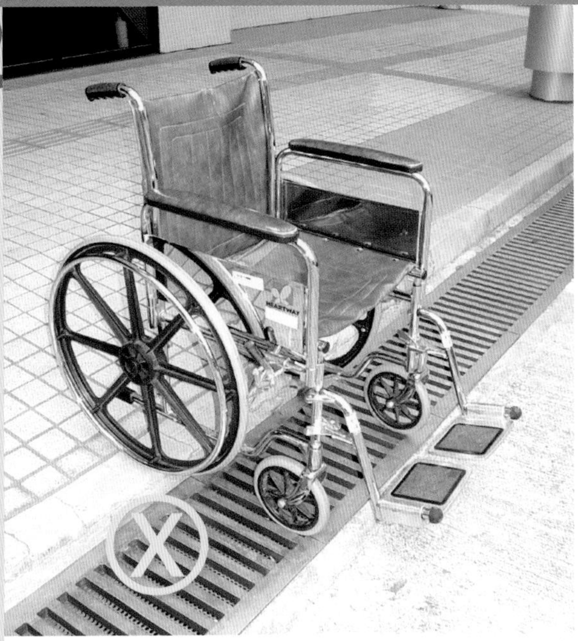

The openings for the gratings are not small enough and are also in the direction of major traffic
渠道格栅开过大且与主要行走方向平行

Preferred opening size of channel grating slots
渠道格栅的推荐开口尺寸

Overhead clearance for an accessible path
方便易达通道的净空高度

3.2

3.2.3 Tactile Surface in External Areas

- To facilitate use by people with visual impairments, it is recommended that a tactile surface shall be provided to pavements, dropped kerbs and paths giving access to buildings. Ideally, tactile surface shall also be provided in communal areas of residential buildings.
- It is recommended that hazard warning tiles shall be provided on dropped kerbs, platform edges of walking surfaces, in front of push-out doors and lift doors.
- It is recommended that positional tiles shall be provided where there is a change in walking direction.
- It is recommended that hazard warning tiles and positional tiles shall be provided to indicate locations to stop for action, such as pedestrian crossings, lift call buttons, emergency call buttons and signboards.
- It is recommended that hazard warning tiles shall be provided to indicate level changes.
- It is recommended that hazard warning tiles shall be provided at the position where caution is required, such as swing-out doors and street crossings.

3.2.3 室外引路径

- 为方便视障人士使用，行人路、下斜路缘石及楼宇进出通道应设有引路径，最好也在住宅建筑内的共用地方设置。
- 建议在下斜路绿石、行人平台边、推门及升降机前设置危险警告引路砖。
- 建议在改变步行方向的位置铺设方位引路砖。
- 建议铺设危险警告和方位引路砖，显示应停下来的位置，例如在行人过路处、升降机呼叫按钮、紧急事故按钮及指示牌的位置。
- 建议应铺设危险警告引路砖，显示地面标高的改变。
- 建议在需要提防的位置应设有危险警告引路砖，例如门开启方向和过街处。

3.2.4 Lighting in External Areas

- It is recommended that the illumination of external areas shall be not less than 50~75 lx. Ideally, it shall be 75~120 lx.

3.2.4 室外照明

- 建议室外光线至少要有50 ~ 75lx，最好有75 ~ 120lx。

Position tiles in front of a signboard
指示牌前的方位引路砖

3

3.3 CAR PARKS FOR BARRIER-FREE ACCESS

3.3.1 Drop-Off Bays / Pick-Up Bays

- It is recommended that drop-off / pick-up bays shall be provided near building entrances close to the main circulation areas or lift lobbies.

3.3.2 Parking in Car Parks

- It is recommended that accessible parking bays shall be provided, close to accessible entrances. The locations of accessible parking bays shall be clearly signed or indicated at the entrance of a car park. Bays and transfer zones shall be marked in yellow for easy identification.

- Allocation of accessible parking space recommended as follows:

Total No. of Car Parking Spaces	Required No. of Accessible Parking Space
1-50	1
5-150	2
151-250	3
251-350	4
351-450	5
Above 450	6

Note: For residential buildings particularly intended for the elderly and people with disabilities, it is recommended that the number of accessible parking bays shall be greater than the above allocations.

3.3 无障碍停车设施

3.3.1 落客处/上客处

- 建议落客/上客处应靠近紧连主要的室内交通通道或升降机大堂的楼宇入口。

3.3.2 停车场泊车设施

- 建议在无障碍入口附近设置停车位，并于停车场入口清楚标示无障碍停车位的位置。停车位和轮椅通道应用黄色圈画，容易识别。

- 无障碍停车位数目的比例建议如下:

停车场车位总数	应设无障碍停车位数目
1~50	1
5~150	2
151~250	3
251~350	4
351~450	5
450 以上	6

注: 为长者和残障人士而设的住宅楼宇，建议无障碍停车位的数目应多于以上的配额。

A drop-off bay close to the entrance of a residential building
住宅建筑物入口的落客处

3.3.3 Transfer Zones for Accessible Parking Bays

- It is recommended that accessible parking bays shall have a transfer zone of at least 1200mm wide marked at the rear of the space or between two parking bays.
- It is recommended that yellow cross-hatching shall be clearly marked on the ground to indicate the transfer zones.

3.4 RAMPS FOR BARRIER-FREE ACCESS

3.4.1 Dimensions for Ramps

- It is recommended that ramps to building entrances shall be provided in a convenient location so that ramp users are not segregated from stair users.

Transfer zones for accessible parking bays
方便易达停车位的转移区

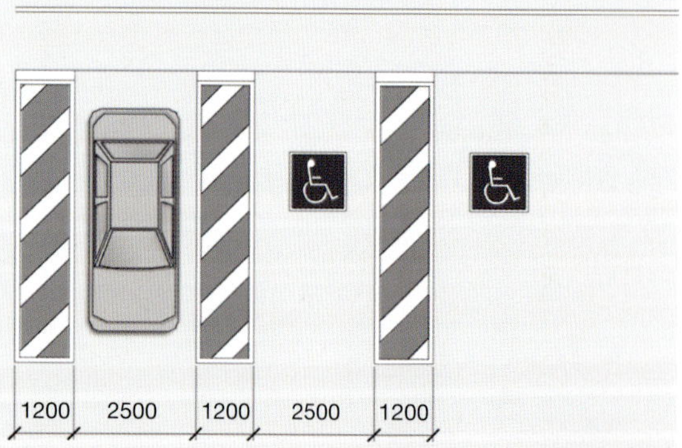

3.3

3.3.3 无障碍停车位的轮椅通道

- 在无障碍停车位的临近车位旁或两个停车位之间，建议保留至少宽1200mm的空间作为轮椅通道。
- 轮椅通道建议在地面上以黄色交叉线标明。

3.4

3.4 无障碍斜道

3.4.1 斜道的尺寸

- 建议通往楼宇入口的斜道应设于方便使用的地点，不应将斜道使用者和梯级使用者分隔开。

A ramp close to the building entrance
建筑物入口的斜道

3.4

- It is recommended that a ramp shall be provided where there is a change in level more than 200mm, particularly if lift services are not available.
- It is recommended that ramps shall be provided at a gradient not steeper than 1 in 12.
- The minimum clear width of a ramp should be 1050mm.
- No ramp should be curved.

- 如地面标高度的改变多于200mm，尤其是如果没有电梯服务，则建议设置斜道。
- 斜道的坡度建议不应大于1:12。
- 斜道的净宽度最少要有1050mm。
- 斜道不应弯曲。

3.4.2 Handrails and Upstands for Ramps

- It is recommended that handrails shall be provided on both sides of all ramps.
- Design of handrails is discussed in Chapter 3.6.
- A raised upstand not less than 100mm high should be provided on any open side of a ramp or a landing.
- Where a ramp with a rise greater than 200mm discharges down to an area where vehicular traffic is possible, it should be provided with a railing or barrier across the full width of its lower end, not less than 1500mm from the lower end.

3.4.2 斜道的扶手及安全挡台

- 建议斜道两边应设有扶手。
- 扶手的设计详见于第3.6节。
- 在任何斜道或斜道平台开放的一端，应设置不少于100mm高的安全挡台。
- 如斜道升高多于200mm，并伸往下方连接马路，则须于斜道末端设置与斜道宽度相等的栏杆或护栏，装置距离斜道末端不可少于1500mm。

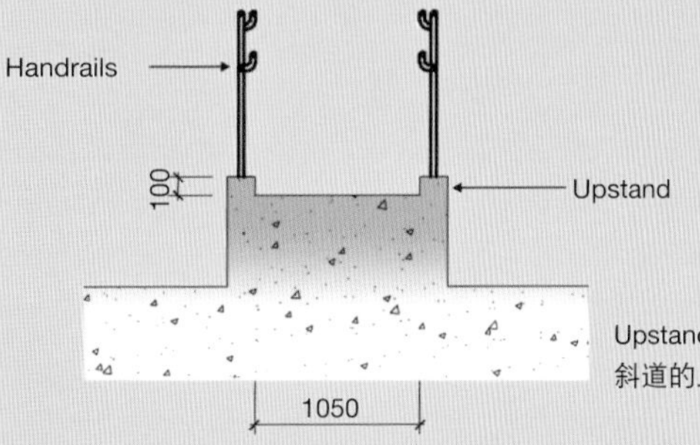

Upstands for a ramp
斜道的上斜竖柱

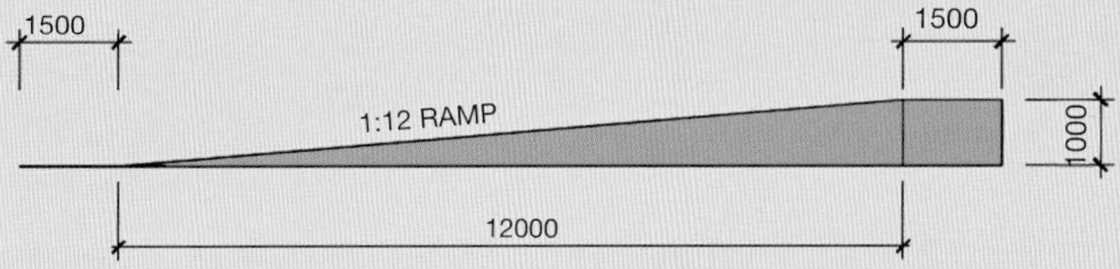

A ramp with a gradient not steeper than 1:12
斜道的坡道不大于1:12

3.4

3.4.3 Landings for Ramps

- It is recommended that level landings not less than 1500mm x 1500mm shall be provided at the bottom and top of a ramp, in all cases clear of any door swing.
- It is recommended that level landings, not less than 1500mm long and the full width of a ramp shall be provided at every 10m.
- Hazard warning tiles should be provided at landings. It is recommended that hazard warning tiles shall be provided 300mm from the top and bottom edges of ramps.

3.4.3 斜道平台

- 斜道顶端及底部建议设有平台，面积不少于1500mmx1500mm，并且此空间内不得有门开入。
- 斜道每隔10m处建议设置长度不少于150mm，宽度与斜道宽相同的平台。
- 斜道平台应设置危险警告引路砖。建议危险警告引路砖应离开每段斜道顶部与底部均300mm。

Hazard warning tiles for ramps
斜道的危险警告引路砖

3

3.4.4 Surfaces for Ramps

- The surface of ramps should be slip-resistant.
- To facilitate use by people with visual impairments, it is recommended that tactile warning strips shall be provided at the top and bottom of a ramp.

3.4.4 斜道面

- 斜道表面必须防滑。
- 为方便视障人士使用，建议在斜道顶部及底部两端须铺设危险警告引路砖。

3.4.5 Visual Factors for Ramps

- To facilitate use by people with visual impairments, it is recommended that contrasting colours shall be provided for floors and walls along ramps.
- Good illumination shall be provided on ramps. It is recommended that an illumination level shall be provided of at least 120 lx, and ideally 300 lx, at the ramp level.

3.4.5 斜道的视觉因素

- 为方便视障人士使用，建议斜道地面和沿途的墙壁要选用对比色。
- 斜道要有充足照明，建议照明度最少要有120lx，最理想照度为300lx。

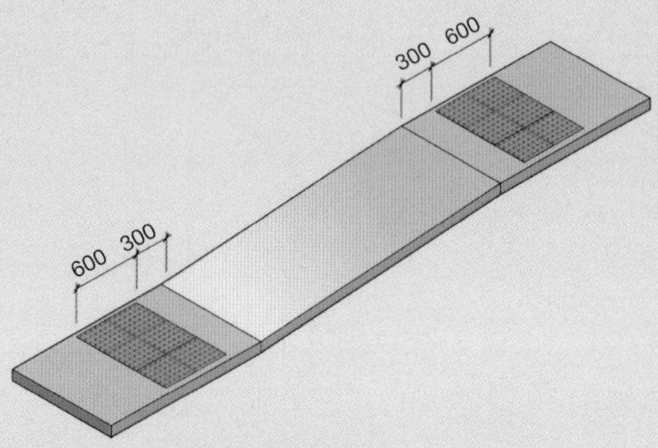

Location of hazard warning tiles at ramps
斜道的危险警告引路砖的位置

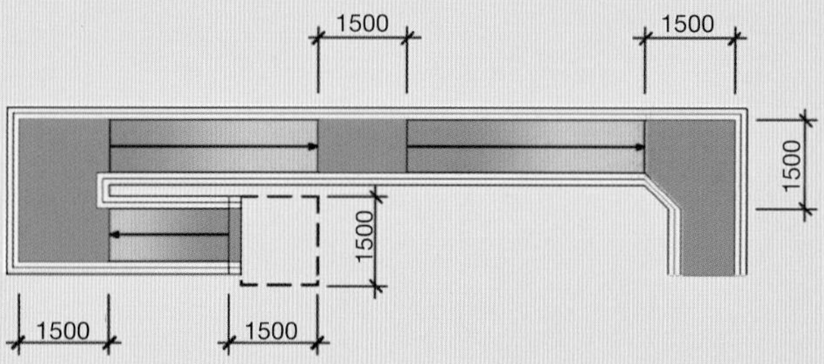

Landings for a ramp
斜道上落处

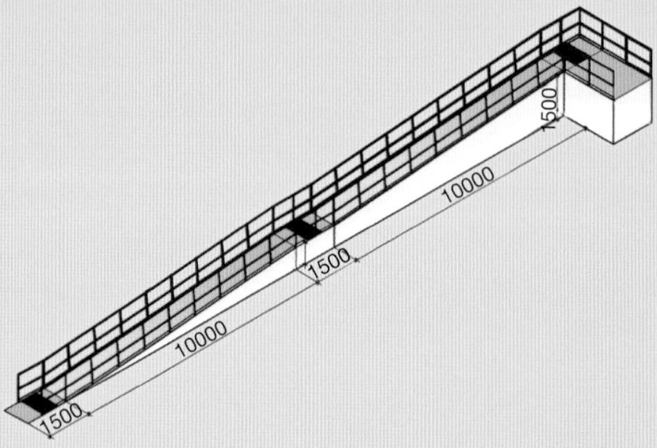

Level landings for a ramp at every 10m
每10m处设于斜道上落处

3.5

3.5 STEPS AND STAIRCASES FOR BARRIER-FREE ACCESS

3.5.1 Dimensions for Steps and Staircases

- It is recommended that for external steps, the depth of treads shall be not less than 280mm and the height of risers shall be not more than 150mm.
- It is recommended that for internal steps, the depth of treads shall be not less than 240mm (measured at the centre of the flight) from the face of one riser to the face of the next riser, and the height of risers be not more than 175mm.
- It is recommended that risers shall be vertical or with a receding face not exceeding 15mm from the vertical, without a projecting nosing.
- Minimum clear width of steps or staircases should be 1050mm.

3.5.2 Landings for Steps and Staircases

- A single flight of stairs should not have more than 16 steps.
- Level landings should be provided at the top and bottom of each flight, with slip-resistant floor surfaces.

3.5 无障碍梯级和楼梯

3.5.1 梯级和楼梯的尺寸

- 建议户外梯级面宽度不应少于280mm，台阶高度不多于150mm。
- 建议室内梯级面宽度，不应小于240mm（由梯级中心点量度），梯级高度不应高于175mm。
- 建议梯级踢面应垂直，或由垂直面往外倾不超于15mm，梯级边缘不得突出。
- 梯级或楼梯间的净宽度至少要有1050mm。

3.5.2 梯级和楼梯的平台

- 每段楼梯不应多于16级梯级。
- 每段楼梯的顶部和底部两端应有平台，并铺设防滑地面。

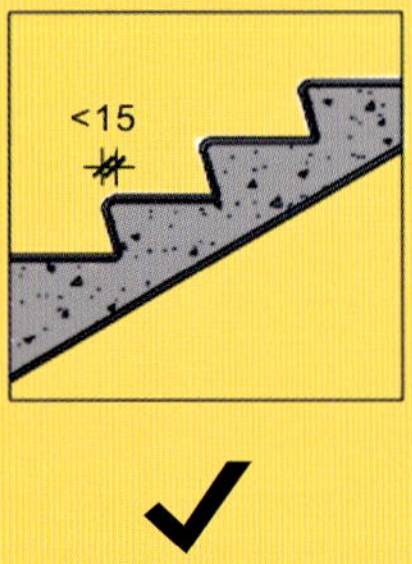

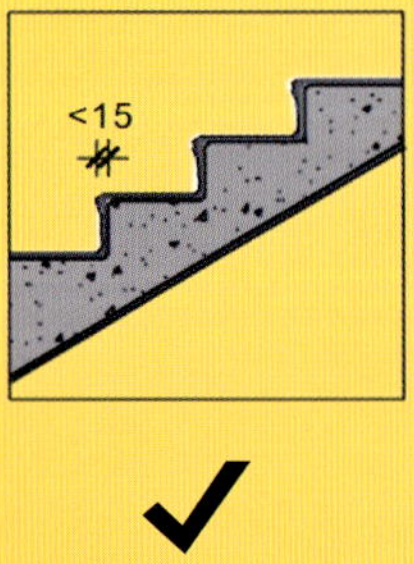

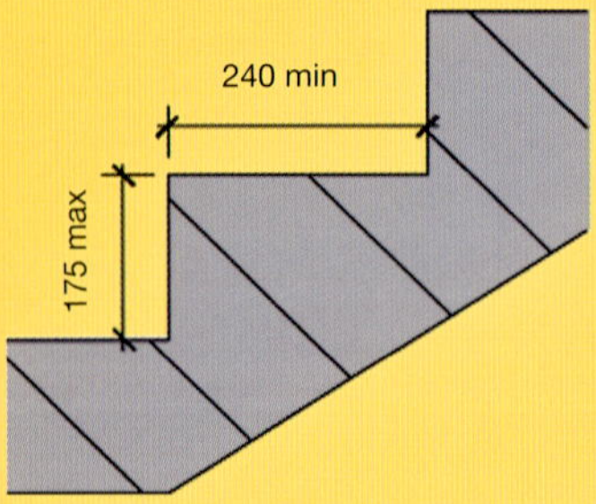

Dimension requirements for internal steps
室内梯级尺寸要求

3

3.5.3 梯级和楼梯的危险警告引路砖

- 为方便视障人士使用，所有平台均应设有危险警告引路砖。
- 危险警告引路砖应离开每段楼梯顶部与底部均300mm。

3.5.4 梯级和楼梯的扶手

- 建议在梯级和楼梯两旁，一律设置不中断的扶手。
- 扶手的设计详见第3.6节。

3.5.5 梯级和楼梯设计的视觉因素

- 为方便视障人士使用，建议楼梯面和墙壁的颜色及色调采用对比色或选用有对比质感的物料。
- 为突出每个楼梯面的边缘，建议铺设有对比色的防滑突缘条，宽度最少要有40mm，离开楼梯面前边缘不多于10mm。
- 所有楼梯面的照明度，建议最少要有120lx。

3.5.6 梯级和楼梯的其他因素

- 不应设置无踢面梯级。
- 不应设置孤形扇形的梯级。

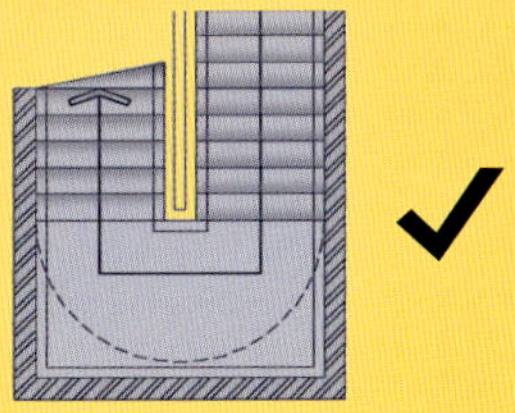

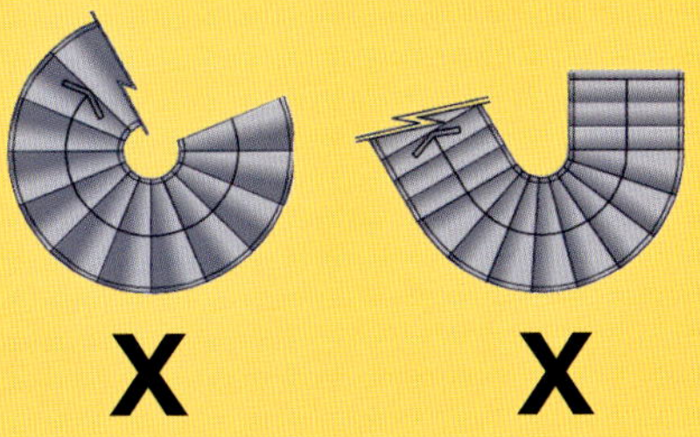

Winders and splyed steps shall be avoided
不应设置孤形扇形的梯级

3.5.3 Hazard Warning Tiles for Steps and Staircases

- To facilitate use by people with visual impairments, hazard warning tiles should be provided at all landings.
- It is recommended that tactile hazard warning tiles shall be provided 300mm from the first and last nosings of every flight of steps.

Hazard warning tiles at landings
楼梯平台上的危检警告引路砖

3.5.4 Handrails for Steps and Staircases

- It is recommended that all steps and staircases shall have continuous handrails installed on both sides.
- Design of handrails is discussed in Chapter 3.6.

3.5.5 Visual Factors for Steps and Staircases

- To facilitate use by people with visual impairments, it is recommended that the treads and walls of a staircase shall be in contrasting colours, tones or materials.
- In order to highlight the edges of individual treads, it is recommended that slip-resistant nosing strips in contrasting colours shall be provided at least 40mm in width and not more than 10mm from the front edge of the tread.
- It is recommended that illumination of not less than 120 lx should be provided at all stair treads.

3.5.6 Other Factors for Steps and Staircases

- Open risers should not be provided.
- Winders and splayed steps should not be provided.

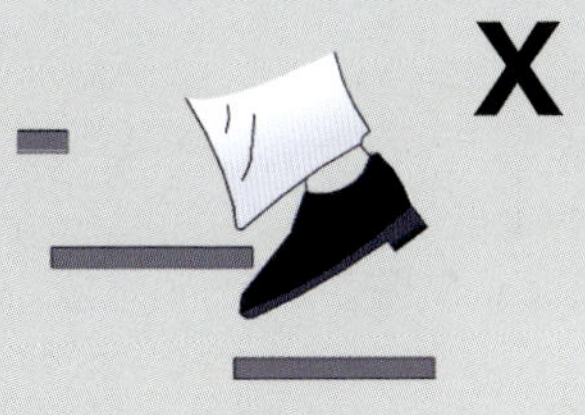

Open risers can cause fall
无踢面梯级会导致行人跌倒

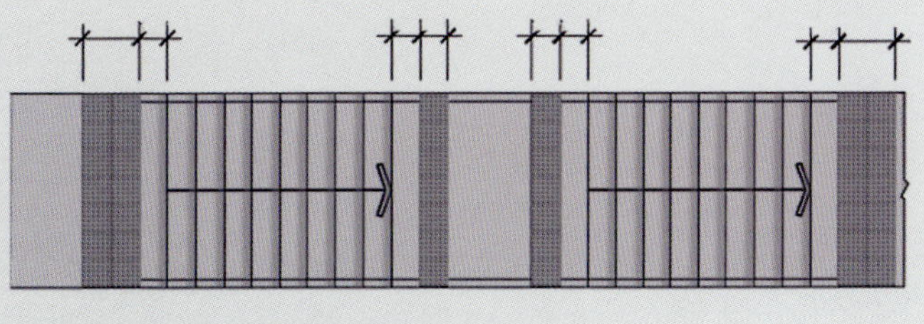

Location of hazard warning tile for stairs
在楼梯处设置危检警告引路砖的位置

3.6 HANDRAILS

3.6.1 Height for Handrails

- It is recommended that handrails shall be provided with their top surface at a height of not less than 850mm and not more than 950mm above the surface of ramps, nosings, floors or landings.

3.6.2 Extension of Handrails

- Handrails should extend horizontally at least 300mm beyond the first and last nosings of every flight of steps, or beyond the end of ramps.
- It is recommended that notes in braille and raised letters shall be provided on the extended parts of handrails to indicate direction or floor level.

Braille on the extension of handrails
扶手延伸部分的凸起盲文

- It is recommended that handrails shall be provided with terminations to prevent injury or catching clothes. There are many ways to achieve this. Options include:
  (a) Turning the end down at least 100mm;
  (b) Returning the end to the wall;
  (c) Returning the end to the last post of a handrail.

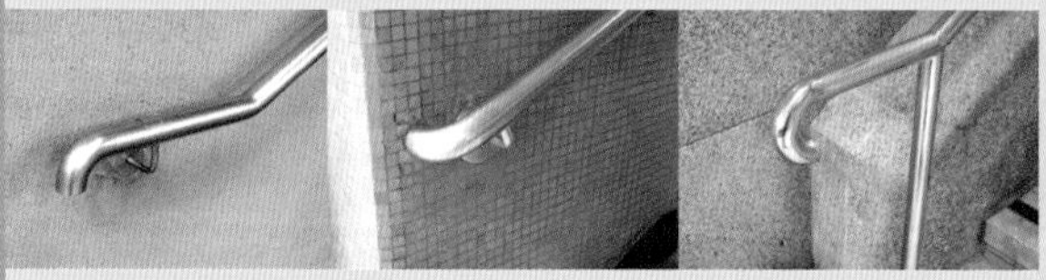

(a) (b) (c)

3.6.3 Shape of Handrails

- Handrails should be comfortable to use by people with impaired dexterity. To achieve this, it is recommended that tubular handrails shall be provided with external diameter not less than 32mm and not more than 40mm.
- It is recommended that a clear space of between 30mm and 50mm shall be provided between the handrails and the walls. .

3.6.4 Loading for Handrails

- It is recommended that handrails shall be provided to resist a load of not less than 1.3 kN applied vertically or horizontally.

3.6 扶手

3.6.1 扶手高度

- 建议扶手顶部离斜道、梯级边、地面或平台面，应介乎850 ~ 950mm。

3.6.2 扶手延伸部分

- 扶手末端应有延伸部分，延伸部分的末端离开每段楼梯的第一个和最后一个梯级边缘至少300mm。
- 在扶手延伸部分建议设有凸起的盲文和字母提示方向或层数。
- 建议扶手的末端应有收尾处理以避免引致伤人或勾到衣物。方法有很多，包括下列选择:
  (a) 末端往下转至少100mm;
  (b) 末端转向墙面;
  (c) 末端转向扶手的最后一根栏杆竖杆。

3.6.3 扶手形状

- 扶手应能让肢体有障碍的人士舒适地使用。建议设置管状扶手，外围直径介乎32 ~ 40mm。
- 扶手与墙壁的净距离建议应介乎30 ~ 50mm。

3.6.4 扶手承载量

- 扶手的纵向及横向的承载量，建议不应少于1.3 kN。

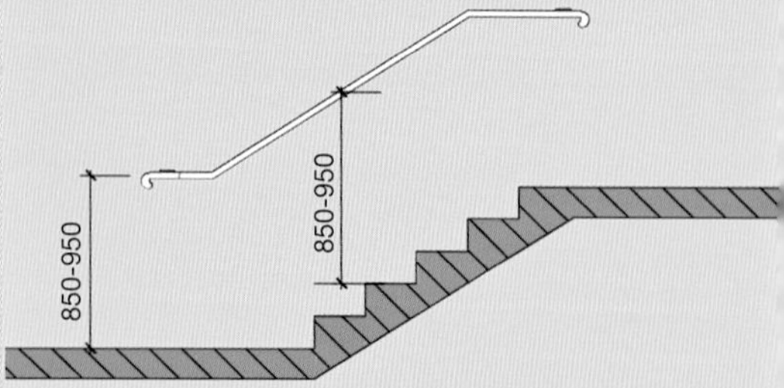

Height of handrails
扶手高度

3.7

3.7 无障碍入口

3.7.1 入口通道

- 建议入口通道的最少净宽度为850mm，理想净宽度为1500mm，以方便两部轮椅同时通过。
- 建议轮椅操作空间平行于门方向的宽度，应为门的总宽度再加上门边至少500mm。
- 入口内外两旁建议提供1500mmx1500mm作为轮椅操作空间。
- 最好可于主要入口及公众的共用地方设置半自动推拉门。
- 主要入口不应设置旋转门，除非旋转门的每一隔足以让一部轮椅或手推车通过。如有旋转门，应于旁边设置另一无障碍的门。
- 门的设计详见于第3.10节。
- 建议在入口处装设闭路电视。
- 建议在入口处装设紧急事故按钮。
- 建议在入口处设置指示牌显示楼宇名称或门牌，字体大小要切合视障人士的需要，并用对比色。

3.7.2 入口平台

- 建议在入口处外设置平台，净深度不少于1500mm。
- 建议在入口平台加建雨棚。

3.7 ENTRANCE FOR BARRIER-FREE ACCESS

3.7.1 Entrance Doorways

- It is recommended that the minimum width of an entrance doorway shall be 850mm, ideally 1500mm to facilitate access of two wheelchairs.
- It is recommended that manoeuvring clearances parallel to an entrance doorway shall include the full width of the doorway plus 500mm minimum.
- It is recommended that a manoeuvring area of 1500mm x1500mm shall be provided on both sides of an entrance door.
- Ideally, semi-automatic sliding doors can be provided at main entrances and in public communal areas.
- It is recommended that revolving doors should not be provided as a main entrance unless the door compartment is sufficiently large to accommodate a wheelchair or a pushchair. Where there is a revolving door, an accessible alternative doorway should be provided adjacent to it.
- Design of doors is discussed in Chapter 3.10.
- It is recommended that CCTV shall be provided at entrances.
- It is recommended that emergency call buttons be provided at entrances.
- It is recommended that a sign indicating the building name or block number should be provided at the entrance, with letters / characters in sufficient size and in contrasting colours to assist people who have visual impairments.

3.7.2 Entrance Platforms

- It is recommended that a platform shall be provided outside an entrance, with a clear width of not less than 1500mm.
- It is recommended that overhead canopy shall be provided above the entrance platform.

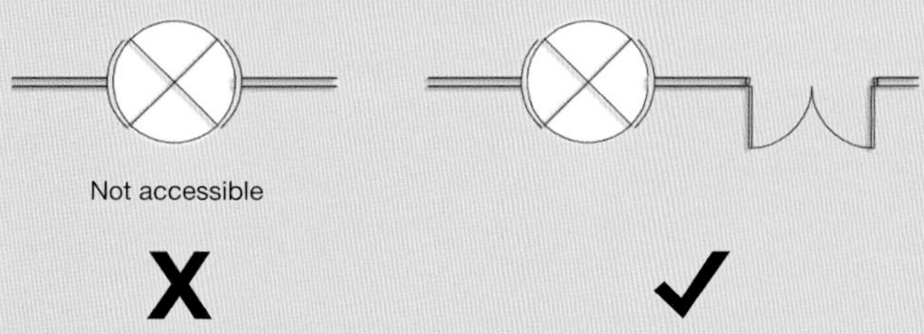

Revoiving door
旋转门

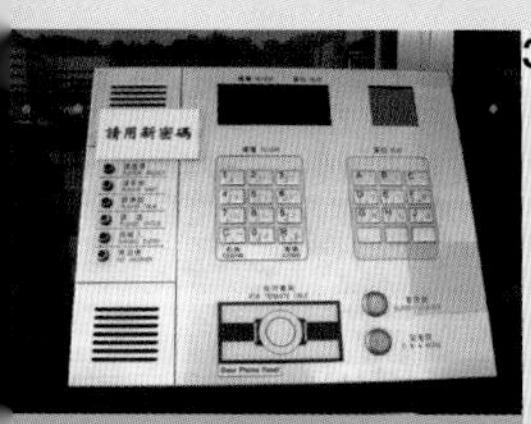

Intercom panel at building entrance
入口处的对讲机

3.7

3.7.3 Entrance Platform Lifts

- It is recommended that where an entrance platform is provided and it is impractical to provide a ramp to deal with changes in level, a platform lift or wheelchair stair lifts shall be provided.

3.7.4 Lighting and Communication Equipments for Entrances

- To enable people with visual impairments to locate entrances, sufficient lighting should be provided.
- To facilitate use by people in wheelchairs, it is recommended that security intercom panels, entry phones, door bells or call buttons outside the entrance shall be located close to the entrance door and at a height of between 900mm and 1100mm above floor level, and should be in contrasting colour to facilitate use.
- To assist people with visual impairments, it is recommended that directional and positional tactile surfaces shall be provided to indicate the location of the security intercom panels and entry phones.

3.7.5 Headroom for Entrances

- It is recommended that clear headroom of at least 2000mm shall be provided in all rooms, corridors, passageways, aisles, or other circulation areas.
- To assist people with visual impairments, it is recommended that where vertical clearance of an area adjoining an accessible route is reduced to less than 2000mm, a barrier or warning tiles shall be provided to alert users and prevent traffic.

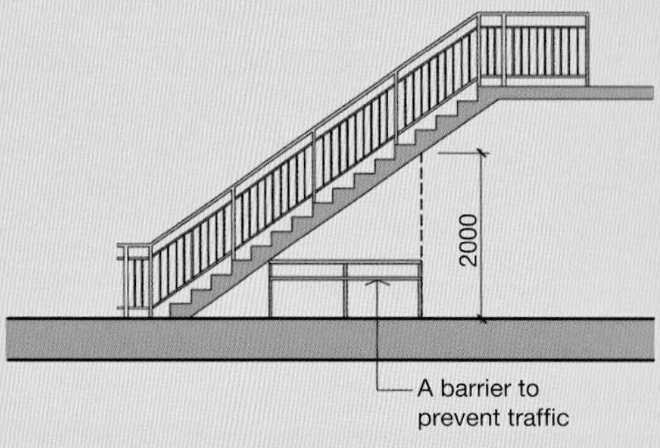

3.7.3 入口平台升降机

- 如入口平台无法设置斜道，则建议设置平台升降机或其他轮椅用楼梯升降机。

3.7.4 入口的照明及通讯设备

- 入口照明度要让视障人士容易找到。
- 为方便轮椅人士，建议在入口处设置保安内部通讯控制板、大门电话对讲系统、门铃，或者唤叫按钮，高度离地面900～1100mm，均采用对比色，方便使用。
- 为协助视障人士，建议设置引路径，指示保安内部通讯控制板，以及大门电话对讲系统的方向和位置。

3.7.5 入口的净空高度

- 房间、走廊、通道、过道或其他交通通道的净空高度最少要有2000mm。
- 为协助视障人士，如某些地方与无障碍通道相连，而垂直净空高度又低于2000mm，则须设置档杆或危险警告引路砖，以警告行人，并避免行人行入此空间而撞头。

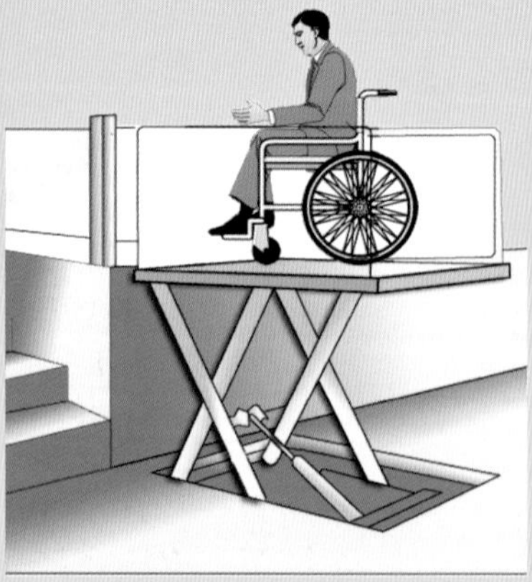

Entrance platform lifts
入口平台升降机

3

Storage area near building entrance
靠近住宅入口处的贮物区

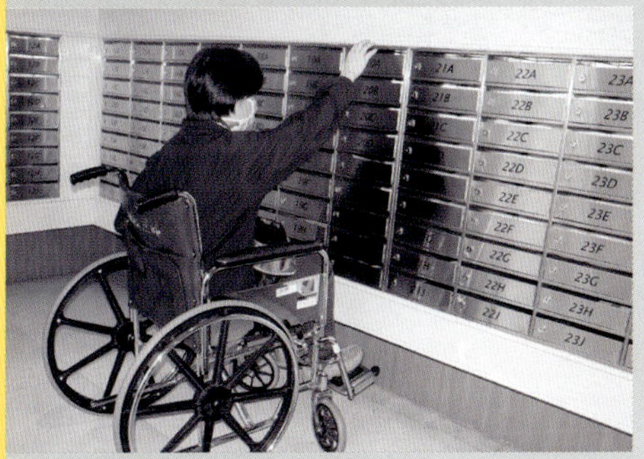
Accessible mailboxes
方便易达的邮箱

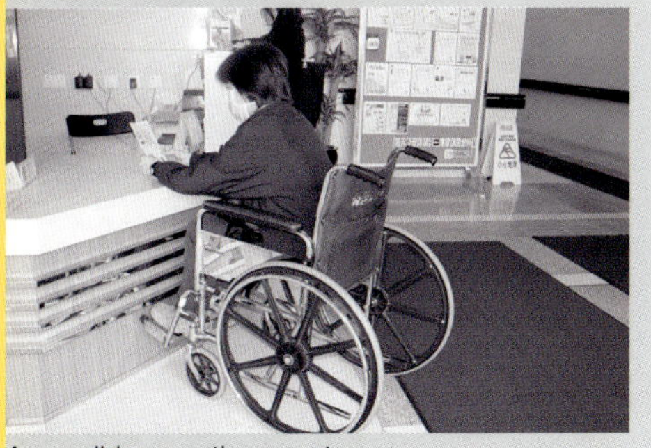
Accessible reception counter
方便易达的接待处

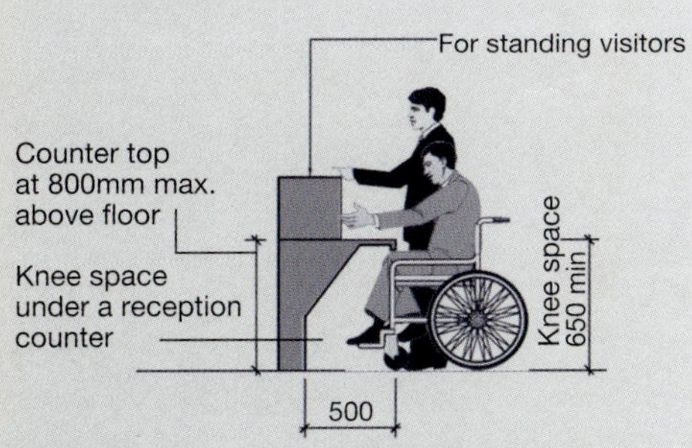

Suggested dimensions for accessible counters
方便易达的接待处的建议尺寸

3.7

3.7.6 Reception Counters for Entrances

- To facilitate use by people in wheelchairs, it is recommended that reception desks shall be at least 1000mm long, with a height between 700mm and 800mm above the floor level and with knee space no less than 500mm deep and 650mm high.

3.7.7 Mailboxes for Entrances

- To facilitate use by people in wheelchairs, it is recommended that the top most mailboxes shall be not more than 1200mm, and the bottom most mailbox shall be not less than 400mm above floor level.
- It is recommended that individual mailboxes shall be identified with raised letters / characters and Braille.

3.7.8 Storage Areas near Entrances

- To facilitate safe and convenient use of habitable areas, it is recommended to provide storage in communal areas for residents' personal property.

3.7.6 接待处

- 为方便轮椅人士，建议接待处柜面至少应有一段长度大于或等于1000mm；而柜面离地面高度介乎700～800mm，且预留膝盖空间的深度和高度不少于500mm和650mm。

3.7.7 邮箱

- 为方便轮椅人士，建议邮箱顶端不应高于1200mm，底部离地面不应低于400mm。
- 建议在每个邮箱设有凸出的盲文和字母/文字。

3.7.8 贮物区

- 为了安全和使用方便，建议在公众的共用地方设置贮物区，方便住客存放个人物品。

3.8

3.8 LIFTS FOR BARRIER-FREE ACCESS

3.8.1 Performance Requirements for Lifts

- It is recommended that access shall be provided to all floors of a building by at least one accessible lift with universally designed features.

3.8.2 Dimensions for Lifts

- It is recommended that the minimum internal dimensions of a lift car shall be 1400mm in depth and 1200mm in width, and ideally the internal dimensions shall be more than 1500mm x 1500mm to accommodate a turn of a wheelchair.
- It is recommended that lift door clear width shall be not less than 850mm.
- It is recommended that lift lobby clear depth shall be not less than 1.8m.

3.8.3 Handrails for Lifts

- It is recommended that handrails shall be provided at the rear and at the sides of a lift car.
- Design of handrails is discussed in Chapter 3.6.

3.8.4 Mirrors in Lifts

- To facilitate use by people in wheelchairs, it is recommended that a mirror shall be provided at the rear of a lift car. The bottom of the mirrors should be not higher than 750mm above the floor of the lift car and the top shall reach the lift ceiling.

3.8 无障碍升降机

3.8.1 升降机设置要求

- 建议在楼宇的每一层设置至少一部符合通用设计设计元素的无障碍升降机。

3.8.2 升降机的尺寸

- 建议升降机厢内的尺寸最少要有1400mm（深）x 1200mm（宽）。理想的机厢内尺寸应为1500mmx1500mm以上，以容纳轮椅旋转 。
- 建议升降机门的净宽度不应少于850mm。
- 建议升降机大堂净深度不应少于1.8m。

3.8.3 升降机的扶手

- 建议升降机厢两边及后方应设置扶手。
- 扶手的设计详见于第3.6节。

3.8.4 升降机镜

- 为方便轮椅人士，建议在升降机厢后方设置镜子，镜底离机厢地面不应多于750mm，镜顶则应与升降机顶齐平。

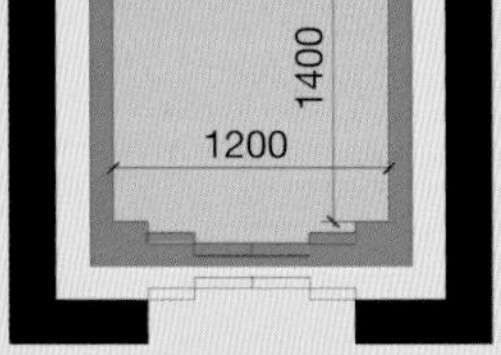

Minimum internal dimensions of a lift car
升降机厢内的最小尺寸

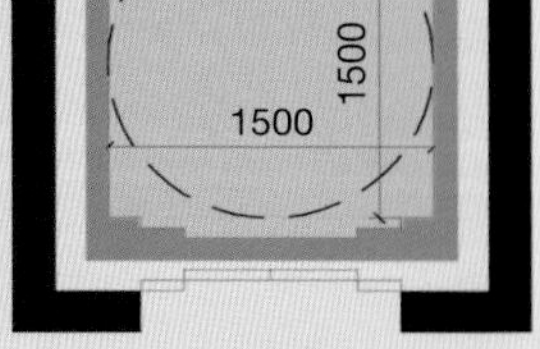

Ideal dimensions of a lift car
升降机厢内的最佳尺寸

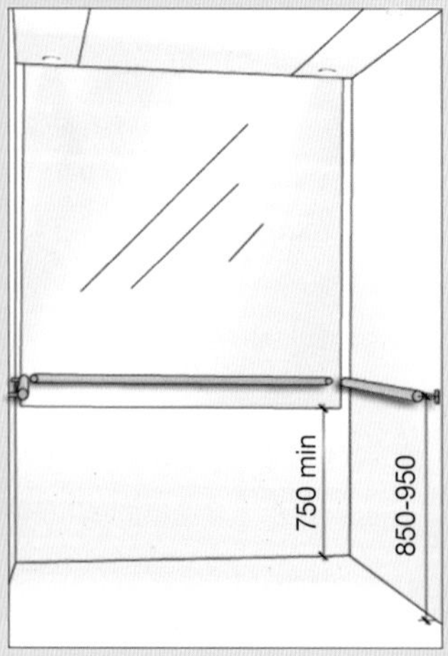

Mirror in a lift
升降机厢内的镜子

3

3.8.5 升降机门

- 升降机门建议应采用横向滑动型。
- 建议升降机门颜色应采用对比色。
- 建议装设发光二极管组开关眼或其他监察装置，不用碰触便可侦测是否有使用者在升降机门中间，令升降机门可及时重开。
- 建议使用广播说明升降机门将要关闭和机厢上落的方向。
- 升降机厢到达每层时，应广播层数。
- 如有人按钮要乘搭升降机，升降机门应保持开敞最少3s。
- 升降机关门速度应低于0.5m/s。
- 建议在电梯大堂设置视听讯号显示最先到达楼层的升降机。

3.8.5 Lift Doors

- It is recommended that lift doors shall be the horizontal sliding type.
- It is recommended that lift doors shall be in contrasting colours.
- It is recommended that a LED array or other detection device shall be provided to prevent doors from closing and initiate re-opening when there is someone in his path, without having to make physical contact.
- It is recommended that an audible announcement shall be set to warn that the door is about to close, and also to indicate the direction of travel of the lift car, e.g. upwards or downwards.
- It is recommended that floor numbers shall be audibly announced on arrival.
- It is recommended that lift doors shall remain fully open in response to a car call for at least 3 seconds.
- It is recommended that door closing speed of the car door shall not be faster than 0.5 metre per second.
- It is recommended that visual and audible signals shall be provided in lift lobbies to indicate which lift car will be the first to reach that floor.

Lift doors in contrasting colours
颜色分明的升降机门

3.8.6 Lift Control Buttons

- To facilitate use by people with visual impairments, it is recommended to provide Braille and raised letter labels on all lift car control buttons and lift lobby call buttons. The emergency alarm push buttons should be in tactile bell shape.
- It is recommended that lift buttons shall have a dimension of not less than 20mm.
- To facilitate use by people with visual impairments, it is recommended that lift buttons shall be in contrasting colours.
- To facilitate use by people in wheelchairs, it is recommended that at least one set of lift car control buttons or lift lobby call buttons shall be located at a height not less than 900mm and not more than 1200mm above the floor level.
- It is recommended that accessible lifts shall be provided with emergency communication equipment, which shall be able to alert the building management once it is lifted. To facilitate use by people in wheelchairs, it is recommended that communication equipment shall be located at a height of not more than 1100mm above the lift floor surface. It can be installed either on a lift wall or inside a cabinet. If inside a cabinet, the door to the cabinet should open and be labelled with Braille and raised letters / characters. A visual signal shall indicate when the emergency communication equipment is in operation.

3.8.6 升降机按钮

- 为方便视障人士使用，升降机厢和升降机大堂内按钮应设凸起盲文和字母，紧急警报按钮应以凸起钟铃形状显示。
- 建议升降机按钮的尺寸不少于20mm。
- 为方便视障人士使用，建议升降机按钮应采用对比色。
- 为方便轮椅人士使用，建议最少应有一组设于机厢内或大堂升降机的按钮，设置在离地板900 ~ 1200mm之间。
- 建议无障碍升降机必须装有紧急通信设备，其装置启动，楼宇管理处应即获得通知。为方便轮椅人士，其安装高度不得离升降机地面多于1100mm，位置须在升降机墙上或紧急设施柜内。如设置在紧急设施柜内，紧急设施柜门必须打开，并附有凸起盲文和字母/文字标示。此外，在使用紧急通信设备时，应有视讯显示。

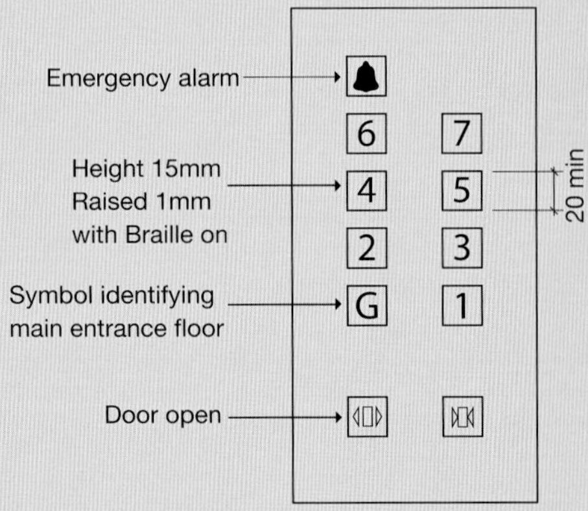

Design of lift control buttons
升降机按钮的设计

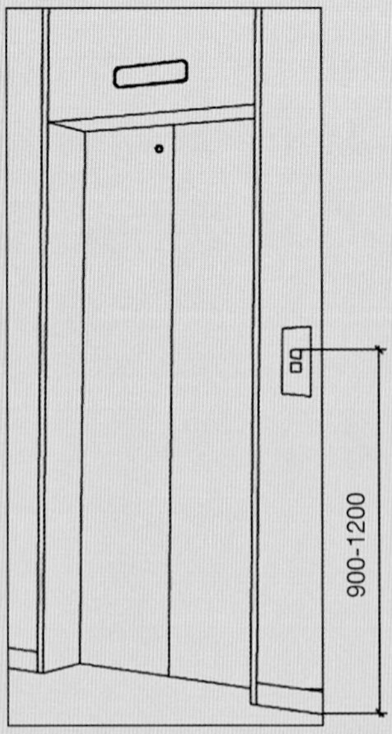

Location of lift call buttons
升降机的呼叫按钮位置

3.8

3.8.7 升降机的地面

- 升降机的地面要防滑。
- 为方便视障人士使用，建议通往升降机门的通道设置引路径。升降机大堂的按钮前建议铺设方位引路砖，升降机前对应处铺设危险警告引路砖。

3.8.8 升降机的座椅

- 为方便长者和体弱人士，升降机内宜设让使用者歇息的座位或长椅。
- 为协助长者和体弱人士，升降机大堂宜设置座椅。

3.8.9 升降机的调平设定

- 建议设置自动调平装置，以确保机厢于规定承载量至零承载量期间，其与地面的距离不多于13mm的公差。

3.8.10 轮椅升降平台

- 建议轮椅升降平台的大小，不应少于1200mmx800mm。
- 轮椅升降平台应设有扶手。扶手设计的详情见第3.6节。
- 为方便视障人士使用，轮椅升降平台的按钮上应设有凸起盲文和字母/文字。

A bench at lift lobby
升降机大堂的座椅

A bench inside a lift
升降机内的座椅

3.8.7 Floor Surfaces for Lifts

- Floor surface for lifts should be slip-resistant.
- To facilitate use by people with visual impairments, it is recommended that a tactile surface shall be provided at access to lift doors. Positional tiles should be provided in front of call buttons in lift lobbies. Hazard warning tiles shall be provided across the front of the lift doors.

3.8.8 Benches for Lifts

- To facilitate use by the elderly and the weak, it is desirable that a perch seat or bench shall be provided inside the lifts.
- To assist the elderly and the weak, it is desirable that seating shall be provided at lift lobbies.

3.8.9 Lift Car Levelling

- It is recommended that self-levelling equipment shall be provided to bring and maintain lift cars at floor landings within a tolerance of 13mm under-rated- loading to zero-loading condition.

3.8.10 Wheelchair Platform Lifts

- It is recommended that the platform of a wheelchair platform lift shall be not less than 1200mm x 800mm.
- Handrails shall be provided at wheelchair platform lifts. Design of handrails is discussed in Chapter 3.6.
- To facilitate use by people with visual impairments, control buttons for platform wheelchair lifts should be labelled in Braille and raised letters / characters.

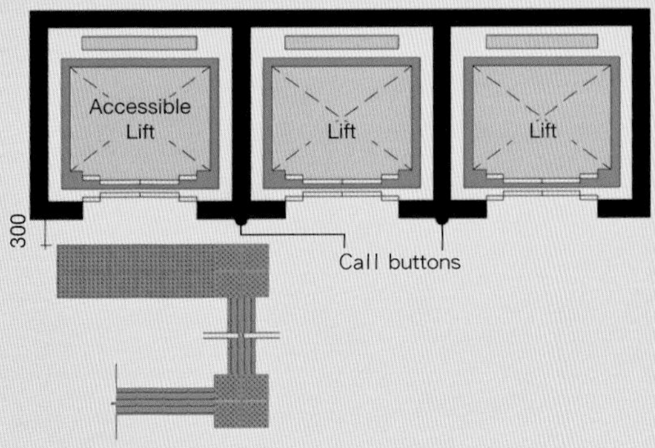

Tactile surface in front of an accessible lift
引路砖设于方便易达的升降机机门前

3.9 CORRIDORS FOR BARRIER-FREE ACCESS

3.9.1 Manoeuvring Space in Corridors

- It is recommended that the clear width of a corridor in a residential building shall be at least 1200mm.
- If a dead-end of a corridor is more than 3500mm in length and less than 1500mm in width, it is recommended that a space of more than 1500mm x 1500mm shall be provided at or within 3500mm of the dead-end.
- Where a door swings out into a corridor, it is recommended that a recessed lobby space shall be introduced. It should be sufficient to contain the door swing and should be not less than 1350mm in width and 900mm in depth.

3.9.2 Floor Surfaces in Corridors

- Slip-resistant floor surface should be provided for corridors.
- Where there is a change in finished floor level of more than 15mm, the joint between the levels shall be splayed.
- To facilitate people with visual impairments to identify the direction and orientation, and also assist in wayfinding, it is preferable that floor patterns be varied at each flat entrance.
- It is recommended that there shall be no projection of more than 100mm from the wall surface and a height of less than 600mm above the floor level.

3.9

3.9 无障碍走廊

3.9.1 走廊的轮椅操作空间

- 住宅楼宇的走廊净宽度建议最少要1200mm。
- 如尽端走廊长度多于3500mm，宽度少于1500mm，则建议在离走廊尽端3500mm内设置大于1500mmx1500mm的空间。
- 如果门扇推向走廊方向，则建议采用凹门廊，宽度不应少1350mm，深度不少于900mm。

3.9.2 走廊地面

- 走廊应铺设防滑地面。
- 地面高差若多于15mm，高差处则要拓成斜面。
- 建议每个居住单位的入口最好铺设不同的地面图案，以帮助视障人士分辨方向及寻路。
- 建议墙上的突出物不得突出墙面超过100mm，其顶端至地面距离应少于600mm。

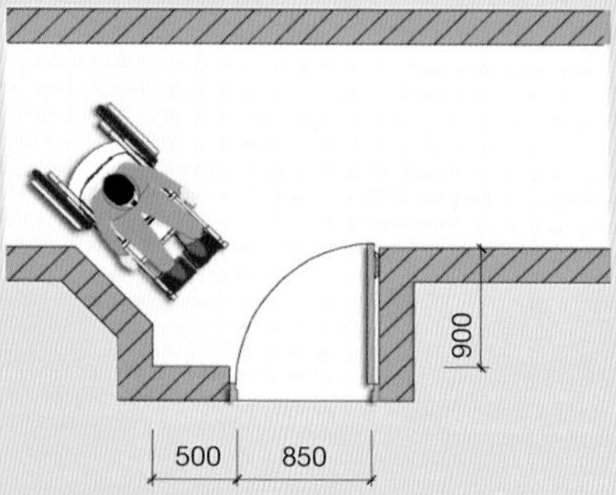

A recessed lobby for a door to swing out into a corridor
隐藏门廊

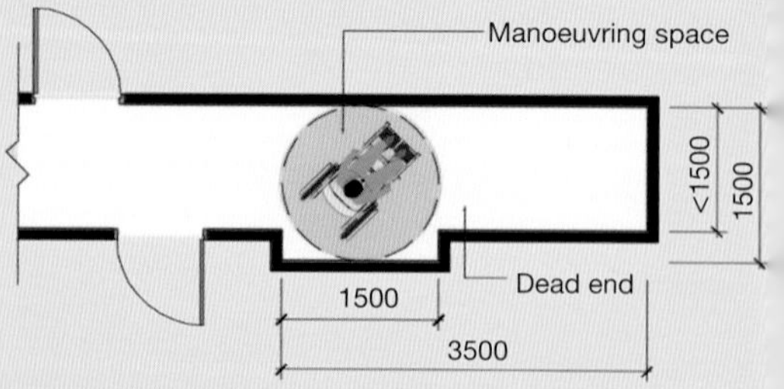

Manoeuvring space in a corridor
走廊上的操作空间

3

3.10

3.10 无障碍门

3.10.1 旋转门和旋转栅门

- 建议无障碍通道上不应装设旋转门或旋转栅门。

3.10.2 门的尺寸

- 建议门的宽度不少于850mm。

3.10.3 门前轮椅操作空间

- 建议轮椅操作空间平行于门方向的宽度，应为门的总宽度再加上门边至少500mm。
- 如连续有多道单向门，建议门与门之间的距离最少有1200mm加门扇宽。
- 如连续有多道双向门，建议门与门之间的距离最少有1500mm加门扇宽。
- 双向自动关闭的门应设有制动装置，防止门扇超过关掩的位置，并设置透明的观望板，底部不得离地面多于1000mm，顶部离地面不少于1500mm。

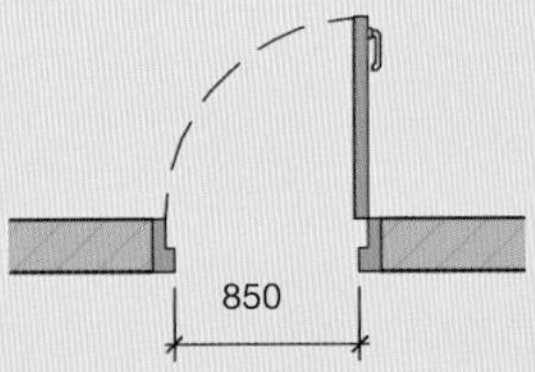

Minimum clear width of a door
门口的最少宽度

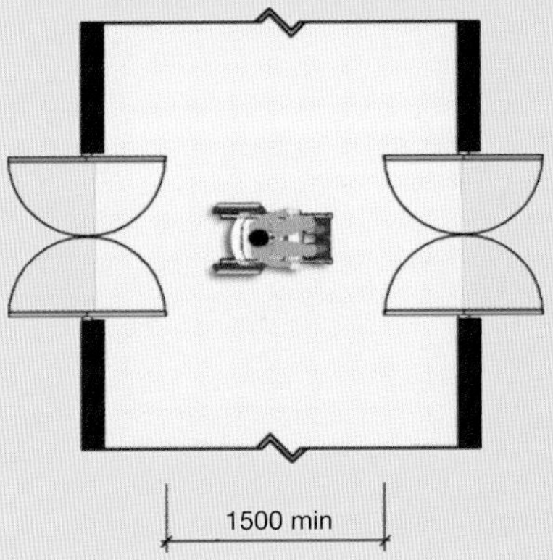

Manoeuvring space for double-action doors
双向门之间的轮椅操作空间

3.10 DOORS FOR BARRIER-FREE ACCESS

3.10.1 Revolving Doors and Turnstiles

- It is recommended that revolving doors and turnstiles shall not be provided on an accessible route.

3.10.2 Dimensions for Doors

- It is recommended that the clear opening width of a door shall not be less than 850mm.

3.10.3 Manoeuvring Space at Doors

- It is recommended that manoeuvring clearances parallel to doorway shall include the full width of the doorway plus 500mm minimum.
- It is recommended that for single-action doors, the distance between doors in series shall be at least 1200mm, plus the width of any door swinging into the space.
- It is recommended that for double-action doors, the distance between doors in series shall be at least 1500mm, plus the width of any door swinging into the space.
- It is recommended that double-action self-closing doors shall be provided with a check mechanism to prevent the doors swinging beyond the closed position. On double-action self-closing doors, a transparent vision-panel shall be provided with a bottom edge not more than 1000mm and a top edge not less than 1500mm above the floor level.

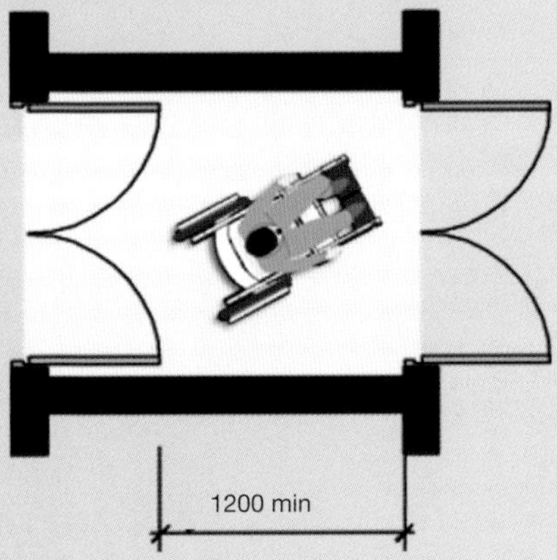

Manoeuvring space for single-action doors
连续单向门之间的轮椅操作空间

3.10

3.10.4 Ironmongery

Door Handles

- It is recommended that door handles be located not less than 950mm and not more than 1050mm above the floor level, measured from the top surface of the grip.
- To facilitate access by people with arthritic hands or limited manual dexterity, it is recommended that lever handles be provided to door latches.
- Where vertical pull handles are provided on doors, it is preferable that the bottom end shall be installed at a height of between 800mm and 1000mm above the floor level.

Keyholes

- It is recommended that at least one keyhole shall be located at a height of not less than 850mm and not more than 1000mm above floor level.

Kickplates

- As wheelchair users sometimes use the footrests of their wheelchairs to open doors, it is recommended to provide kickplates with a height of at least 200mm across the full door width, where wheelchair traffic is expected.

3.10.4 五金器具

门把手

- 由把手上端开始量度，门把手离地面的高度应介乎950～1050mm。
- 建议采用横向条状把手，方便手部关节有毛病或不灵活的人士使用。
- 门户如设有垂直的拉把，底部末端最好安装在离地面800～1000mm高的位置。

匙孔

- 建议最少有一个钥匙孔设于离地面850～1000mm的位置。

踢板

- 由于轮椅人士有时用轮椅搁脚板开门，建议在门户两面可设置至少200mm高的踢板，长度与门户宽度相同。

Lever handles are recommended
建议设置横杆式横向的门把手

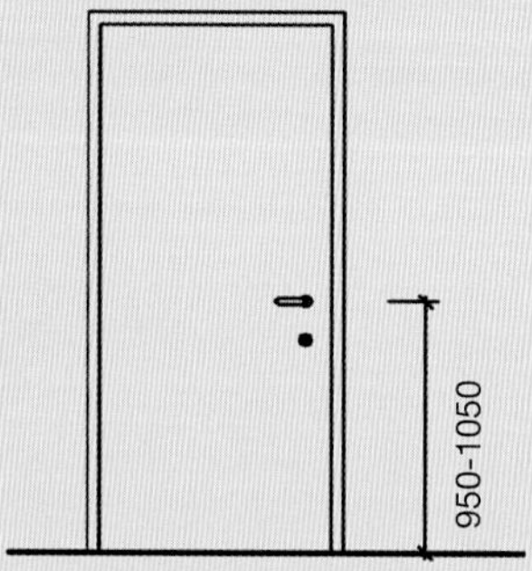

Location of a door handle
门把手的位置

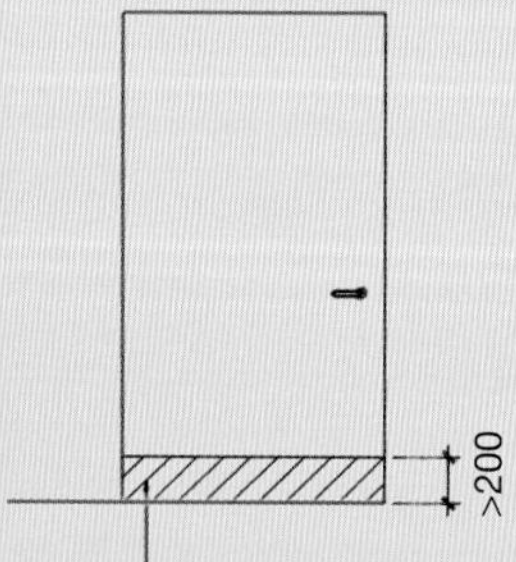

A kickplate for a door
门踢板的位置

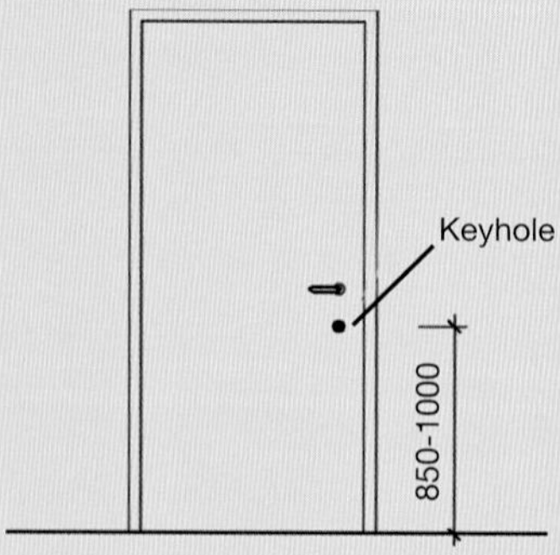

Location of a keyhole
匙孔的位置

3.10

门户开启装置

- 为协助轮椅人士通行，建议半自动门户开启的装置按钮高度应在离开地面1200mm以内，按钮尺寸最少30mmx30mm，并设置凸起的盲文及字母/文字，字体高度不少于10mm。如遇火警或电力中断等紧急事故，半自动门户开启装置应会自动返回开门或人工开关门的设定。
- 键垫装置顶端离地面的高度不得超过1200mm。
- 建议按钮板和键垫上应有凸起的盲文和字母/文字，键垫上“5”字按钮的中央应有凸出1mm的凸点。

门铃

- 为方便听觉障碍人士，建议提供有灯光的视觉门铃或震动式门铃。

3.10.5 门槛

- 门槛高度不可超过15mm，并应切成斜面。

3.10.6 门拉力

- 门户不应过重，如使用沉重的物料，应采用切合门拉力的门户关闭器和五金器具。
- 建议门拉力不应大于25N，方便体弱人士通行。

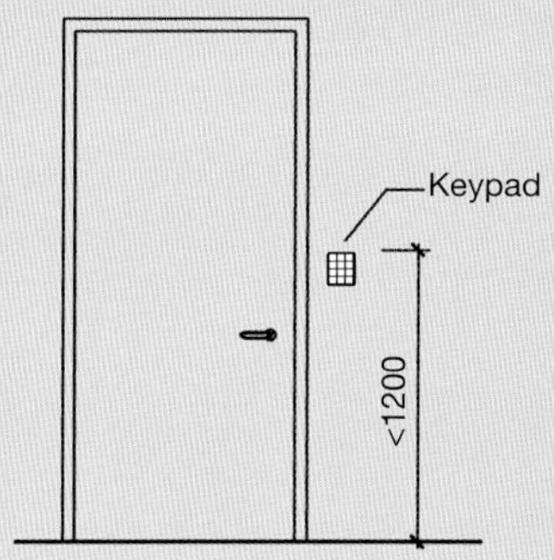

Location of a keypad for a door
门户键垫的位置

Door Openers

- To assist access by wheelchair users, it is recommended that where a semi-automatic door opener is used, the activating button shall be located at a height of not more than 1200mm above floor level, with a press plate of at least 30mm x 30mm in size, labelled in Braille and raised letters / characters not less than 10mm high. In case of emergency, such as fire or power failure, operation of semi-automatic door openers should revert to door open or close.
- Where there is a keypad, the top surface shall be at a height of not more than 1200mm above floor level.
- It is recommended that the letters / characters on press plates or keypads shall be in Braille and raised letters, and a 1mm raised dot shall be provided at the centre of the figure of “5” button of a keypad.

Doorbells

- To facilitate access by people with hearing impairments, it is recommended that visual doorbells with light or vibration indicators shall be provided.

3.10.5 Thresholds

- It is recommended that door thresholds shall not exceed 15mm in height and shall be levelled.

3.10.6 Door Opening Force

- It is preferable that doors shall not be heavier than necessary. Where heavy materials are used, appropriate door closers and ironmongery shall be provided appropriate to the door force.
- To facilitate access by the weak, it is recommended that door opening force shall be not more than 25 newtons.

3.10

3.10.7 Glass Doors

- To facilitate access by people with visual impairments, it is recommended that glass doors and panels shall be provided with clear and colour contrasting markings at a height of between 1200mm and 1500mm to warn people of the presence of the glass.

3.10.7 玻璃门

- 玻璃门和门板应有清晰的标记并采用对比色，高度介乎1200～1500mm，让视障人士知道玻璃的位置。

3.11

3.11 KITCHENS FOR BARRIER-FREE ACCESS

3.11.1 Dimensions for Kitchens

- It is recommended that the clear width of an accessible path in a kitchen should not be less than 1200mm. Where kitchen counters are located on bothsides of the path, the clear width of the accessible path should not be less than 1500mm.
- To facilitate action by wheelchair users, it is recommended that the top surface of kitchen counters be between 750mm and 800mm above the floor level. The depth of the kitchen counters shall be between 500mm and 600mm.
- It is recommended that the bottom of wall-hung kitchen cabinet should be not more than 1200mm above the finished floor level, and the depth shall be not more than 250mm.

3.11.2 Visual Factors

- To facilitate use by people with visual impairments, it is recommended that stoves, cabinets and walls shall be in contrasting colours.
- It is recommended that task lighting be provided at food preparation tops and countertops.

3.11 无障碍厨房

3.11.1 厨房尺寸

- 建议在厨房内无障碍通道净宽不应少于1200mm。如通道两旁设有柜台，无障碍通道的净宽度不应少于1500mm。
- 为方便轮椅人士，建议厨房柜台台面离地面的高度应介乎750～800mm之间，柜台深度应介乎500～600mm之间。
- 建议悬挂式厨柜的底部应离地面不少于1200mm，其深度不应超过250mm。

3.11.2 厨房设计的视觉因素

- 为方便视障人士，炉具、厨柜和墙壁应采用对比色。
- 预备食物的工作台和台面要设置工作灯。

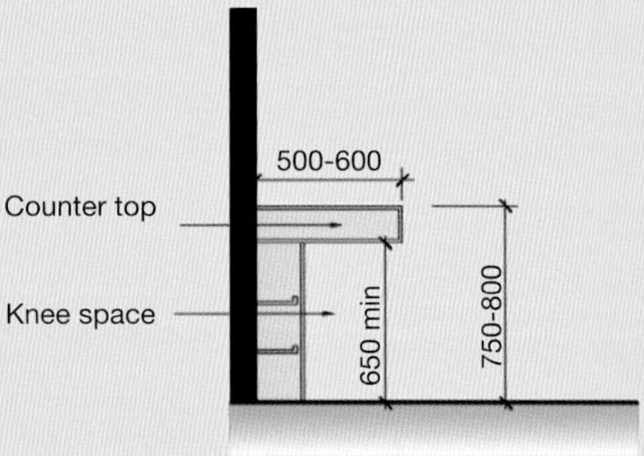

Dimensions of kitchen counters
厨房柜台尺寸

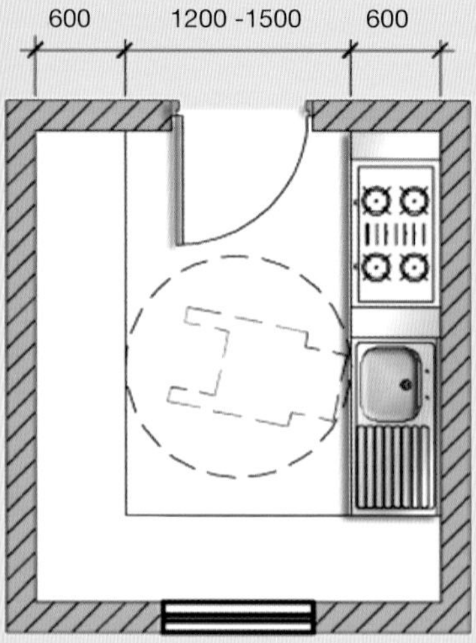

An accessible path in a kitchen
厨房的方便易达通道

3

3.11

3.11.3 厨房内预留膝盖空间

- 建议厨房柜台以下最少应有650mm高的膝盖空间。
- 为方便轮椅人士，建议在炉具、洗涤盆和煮食工作台下面要预留膝盖空间。煮食工作台应设置铰链或可折向墙壁，让厨房有更多空间进行其他活动。

3.11.3 Knee Space for Kitchens

- It is recommended that adequate knee space with a height of at least 650mm shall be provided below the kitchen counters.
- To facilitate use by wheelchair users, it is recommended to provide knee spaces under stoves, sink units and food preparation tops. To allow the space to be used for other activities, the food preparation tops could be either hinged or folded on the wall.

3.12

3.12 无障碍厕所、浴室及淋浴间

3.12 TOILETS, BATHROOMS AND SHOWERS FOR BARRIER-FREE ACCESS

3.12.1 厕所、浴室及淋浴间地面

- 建议铺设防滑地面。

3.12.1 Floor Surfaces in Toilets, Bathrooms, and Showers

- It is recommended that slip-resistant floor finishes shall be provided.

3.12.2 厕所、浴室及淋浴间的地面净空间

抽水马桶的地面净空间

- 为方便轮椅人士转移至抽水马桶，建议抽水马桶地面净空间宽度至少为1500mm，深度由后方墙面计起至少1750mm。

3.12.2 Clear Floor Space in Toilets, Bathrooms, and Showers

Clear Floor Space for Water Closets

- To facilitate transfer to a water closet by people in wheelchairs, it is recommended that clear floor space shall be provided of not less than 1500mm in width and 1750mm in depth from the rear wall.

Knee space for the sink and stove
洗涤盆和炉具下的膝部空间

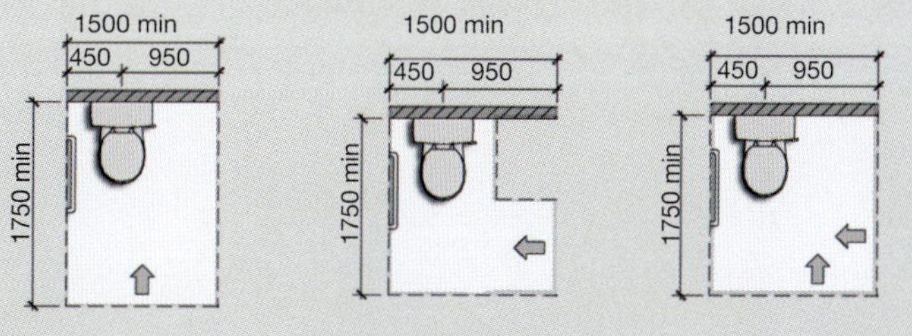

Clear floor space for frontal transfer

Clear floor space for side transfer

Clear floor space for diagonal transfer

Clear floor space for water closets
水厕的地面净空间

Clear Floor Space for Bathtubs and Showers

- To facilitate transfer by people in wheelchairs, it is recommended that clear floor space of at least 800mm x1500mm shall be provided in front of a bathtub or shower.

浴缸及淋浴装置的地面净空间

- 浴缸或淋浴装置前方地面的净空间最少要有800～1500mm，方便轮椅人士转移至浴缸或浴椅。

3.12.3 Wash Basins

- To facilitate use by people in wheelchairs, it is recommended that the rim of a wash basin counter shall be not higher than 800mm above floor level.
- To facilitate use by people in wheelchairs, it is recommended to provide knee space under the washbasins.

3.12.3 洗面盆

- 为方便轮椅人士，洗面盆台面的高度应为从盆边至地面不超过800mm。
- 为方便轮椅人士，洗面盆下方应预留膝盖空间。

3.12.4 Water Closets

- It is recommended that the seat of a water closet shall be at a height of between 450mm and 480mm above floor level.
- It is recommended that flushing controls shall be at a height of between 600mm and 1100mm above floor level. Preferably, lever flushing controls shall be provided.
- It is recommended that flushing controls shall provide adjustable flushing water volume.
- To facilitate use by the weak and the elderly, it is recommended to provide seat lids at water closets. Seats should not be spring-actirated.

3.12.4 抽水马桶

- 抽水马桶坐板面离地面的高度应介乎450～480mm。
- 建议冲水把手离地面的高度应介乎600～1100mm，最好提供杠杆式冲水把手。
- 建议冲水把手应可调校冲水量。
- 为方便体弱人士和长者，抽水马桶应设有座椅盖，抽水马桶不应有弹簧装置。

3.12.5 Urinals

- Mounting height for urinals may vary with predicted use and model characteristics. Where urinals are installed, it is recommended that the front rim shall be at a height of not more than 400mm above floor level.
- To facilitate use by people with visual impairments, it is recommended to provide positional tiles in front of a urinal.
- To assist use by people with ambulant impairments, it is recommended that vertical support rails be provided to at least one urinal, with bottom ends at a height of not more than 600mm and top ends not less than 1200mm.

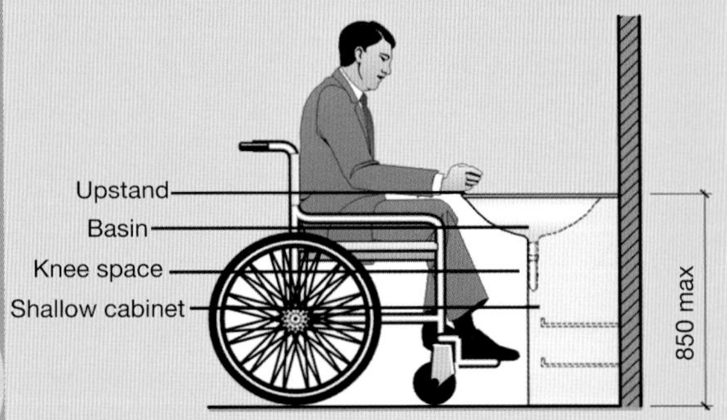

Knee space for a wash basins
洗面盆的膝盖空间

3.12.5 尿厕

- 尿厕的高度或因预计的使用和模式特点而不同，安装尿厕时，建议尿厕前方的盆边离地不应高于400mm。
- 为方便视障人士使用，建议在尿厕前方铺设方位引路砖。
- 为方便行动不便人士使用，建议最少在一个尿厕提供垂直扶手，扶手底部末端离地面不多于600mm，顶端离地面不少于1200mm。

Vertical handrails for a urinal
尿厕的垂直式扶手

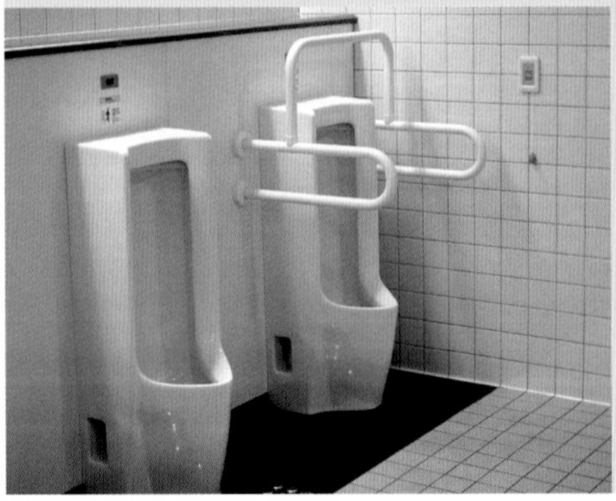

3.12

3.12.6 淋浴装置

- 为方便体弱人士和长者，建议设置淋浴椅。
- 建议浴椅的高度应介乎离地面400 ~ 480mm。
- 建议浴椅表面应该防滑。
- 建议花洒头的高度应可调校，并且花洒头可取下。

3.12.6 Showers

- To facilitate use by the weak and elderly, it is recommended to provide a sturdy seat inside a shower cubicle.
- It is recommended that shower seats shall be at a height of between 400mm and 480mm above floor level.
- It is recommended that shower seats shall be slip-resistant.
- It is recommended that the height of shower heads shall be adjustable in height and be detachable.

3.12.7 浴缸

- 建议浴缸边离地面的高度不应多于300mm。如浴缸边高于300mm，高度不应超过450mm，同时宜使用转移平台或板子，方便行动不便的人士使用。
- 为方便儿童，建议在浴缸间设置防滑梯级。
- 浴缸面应防滑。

3.12.7 Bathtubs

- It is recommended that the rim of a bathtub be at a height of not more than 300mm above floor level. Where the rim of a bathtub is higher than 300mm, the height should not exceed 450mm, and a transfer platform / board should be used to facilitate access by the less agile.
- To facilitate use by children, it is recommended to provide slip-resistant steps at bathtubs.
- Bathtub surfaces should be slip-resistant.

3.12.8 厕所、浴室和淋浴间的扶手

- 扶手的设计详见第3.6节。
- 扶手的净承载量最少应有150kg。

抽水马桶扶手

- 为方便体弱人士和长者，建议至少在抽水马桶的两边各设置一个扶手或在一边设置一个L形扶手。
- 建议最少有一个扶手的安装位置，其末端离抽水马桶中间应有450mm。
- 建议最少设置一个长度不少于600mm的横向扶手，其中一个末端离后方墙面300mm，离地面高度不多于750mm。

3.12.8 Handrails for Toilets, Bathrooms and Showers

- Design of handrails is discussed in Chapter 3.6.
- Handrails shall be capable of supporting a static load of at least 150kg.

Handrails for Water Closets

- To facilitate use by the weak and the elderly, it is recommended to provide at least two handrails to both sides of or one L-shape handrail to one side of a water closet.
- It is recommended that at least one handrail shall be installed with its ends 450mm away from the centre of the water closet.
- It is recommended that at least one horizontal handrail shall have a length of not less than 600mm, with one end 300mm away from the rear wall and at a height of not more than 750mm above floor level.

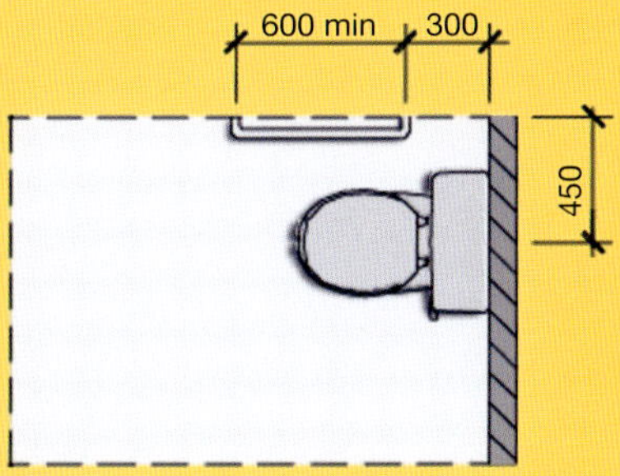

Handrails for a water closet
水厕的扶手

The rim of the bathtub is low for easy access
浴缸边缘较低，容易使用

Handrails for Water Closets (continued)

- It is recommended that a vertical handrail shall have a minimum length of 600mm, with its bottom end at a height of not more than 750mm above floor level.
- It is recommended that the vertical part of a L-shape handrail shall have a minimum length of 600mm, with its bottom end and the horizontal part at a height of not more than 750mm above floor level.

Handrails for Showers

- To facilitate use by the weak and the elderly, it is recommended to provide a horizontal L-shaped handrail, with a length of 750mm on the shower seat wall, from the rear (seat) wall to the adjacent wall, at a height of 750mm from the floor of the shower stall.

Handrails for Bathtubs

- To facilitate use by the weak and the elderly, it is recommended to provide at least one horizontal handrail on the wall adjacent to a bathtub, with a length of not less than 600mm and at a height of between 150mm and 300mm above the bathtub rim.

3.12.9 Water Temperature

- It is recommended that basin taps shall clearly indicate the hot and cold selections, by using contrasting colours.
- It is recommended that water shall not be delivered hotter than 43°C. Thermostatic devices can be installed to control the water temperature.

抽水马桶扶手 (续上)

- 建议垂直扶手的长度最少要有600mm长，底部离地面的高度不应多于750mm。
- 建议 L形扶手的垂直部分的长度最少要有600mm，垂直部分底部和横向部分离地面的高度不应多于750mm。

淋浴间扶手

- 为方便体弱人士和长者使用，建议在浴椅靠边的墙壁设置横向的L形扶手，位置由墙壁后方（浴椅）伸展至邻边墙壁，于浴椅后方墙面上的扶手，长度为750mm，扶手高度离淋浴间地面750mm。

浴缸扶手

- 为方便体弱人士和长者，建议在浴缸邻边墙壁最少设置一个横向扶手，长度不少于600mm，离浴缸顶边的高度应介乎150～300mm。

3.12.9 水温

- 建议水龙头应清楚显示冷、热水，要用对比色区分。
- 建议水温不应高于43℃，并设置控制水温的趋热性装置。

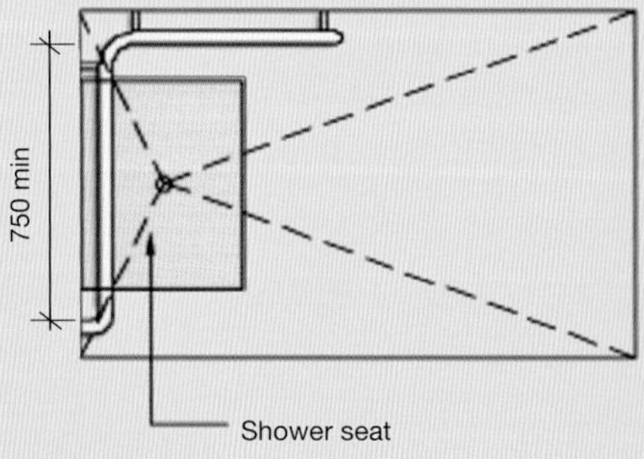

Handrails for a shower
淋浴装置扶手

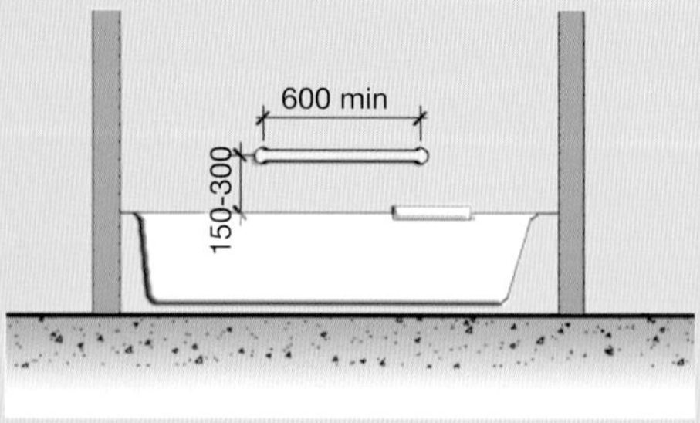

Handrails for a bathtub
浴缸扶手

3.13 无障碍卧室与客/饭厅

3.13.1 卧室与客/饭厅的家具尺寸

- 为方便轮椅人士，建议卧床的硬床垫顶离地面的高度应介乎450～500mm之间。
- 为方便轮椅人士，建议架子高度不应多于1200mm，深度不应多于400mm。
- 镜子位置要切合儿童和轮椅人士的需要，底部离地面不应高于750mm。

3.13.2 卧室与客/饭厅的轮椅操作空间

- 为方便轮椅人士，建议床边通道宽度应不少于900mm，最理想有1500mm，以便轮椅转动和其他人士提供协助。
- 为方便轮椅人士，椅桌旁要备有轮椅操作空间，宽度最少有900mm。
- 为方便轮椅人士，建议通往卧室、客厅、厨房、厕所和贮物房的走廊宽度应不少于1050mm，理想宽度为1200mm。
- 为方便轮椅人士，建议每间卧室和客厅内有1500mmx1500mm的轮椅转动空间。

3.13.3 卧室与客/饭厅的窗户

- 为安全计，离地面900mm内不应设有可开启的窗户。
- 为方便操作，窗户手柄高度不应离地面多于1200mm。
- 如有窗台，窗户的设计应考虑安全性及方便操作，以顾及轮椅人士的需要。

3.13

3.13 BEDROOMS AND LIVING / DINING ROOMS

3.13.1 Furniture Dimensions for Bedrooms and Living / Dining Rooms

- To facilitate use by people in wheelchairs, it is recommended that the height of a bed to the top of the firm mattress shall be between 450mm to 500mm above floor level.
- To facilitate use by people in wheelchairs, it is recommended that shelves shall be provided at a height of not more than 1200mm and with a depth of not more than 400mm.
- To facilitate use by children and people in wheelchairs, it is recommended that mirrors shall be provided with their bottom at a height of not more than 750mm above floor level.

3.13.2 Manoeuvring Space in Bedrooms and Living / Dining Rooms

- To facilitate use by people in wheelchairs, it is recommended that bed aisles shall be provided with a width of not less than 900mm, ideally 1500mm to allow assistance and turning.
- To facilitate use by people in wheelchairs, it is recommended to provide manoeuvring space at tables and seating, with a width of not less than 900mm.
- To facilitate use by people in wheelchairs, it is recommended that corridors to bedrooms, living rooms, kitchens, toilets and stores shall have a width of not less than 1050mm, ideally 1200mm.
- To facilitate use by people in wheelchairs, it is recommended that a turning space of 1500mm x 1500mm shall be provided in every bedroom and living room.

3.13.3 Windows for Bedrooms and Living / Dining Rooms

- For safety, it is recommended that where openable windows are provided, their bottom edge shall be not more than 900mm above floor level.
- For easy operation, window handles shall be not more than 1200mm above floor level.
- Where there is a bay window, safe and easy operation of windows by people in wheelchairs shall be considered.

3.13

3.13.4 Switches and Socket Outlets for Bedrooms and Living / Dining Rooms

- For easy operation, it is recommended that switches and socket outlets shall be mounted at a height of between 900mm and 1100mm above floor level.

3.13.4 卧室与客/饭厅的电器开关与插座

- 为方便操作，建议开关与插座离地面的高度应介乎900～1100mm之间。

3.14

3.14 BALCONIES AND UTILITY SPACES FOR BARRIER-FREE ACCESS

3.14 无障碍露台与工作平台

3.14.1 Dimensions for Balconies and Utility Spaces

- To facilitate use by people in wheelchairs, it is recommended that the depth of a balcony or a utility space shall be not less than 1500mm.
- For safety, the top of a balcony railing shall be at a height of not less than 1100mm.

3.14.1 露台与工作平台的尺寸

- 为方便轮椅人士，露台或工作平台的深度建议不少于1500mm。
- 为安全起见，露台的栏杆顶端离地面至少1100mm。

3.14.2 Manoeuvring Space at Balconies and Utility Spaces

- To facilitate use by people in wheelchairs, it is recommended that a clear manoeuvring space of not less than 1500mm x 1500mm shall be provided.

3.14.2 露台与工作平台的操作空间

- 为方便轮椅人士，建议于露台或工作平台提供不少于1500mmx1500mm的轮椅操作空间。

3.14.3 Laundry at Balconies and Utility Spaces

- It is recommended that an accessible location shall be designated for drying racks, ideally open to the outside air.
- To facilitate use by people in wheelchairs, it is recommended that drying racks shall be installed at a height of not more than 1200mm. Where it is a difficult to install drying racks at a height less than 1200mm, mechanical devices should be provided to lower drying racks for access.

3.14.3 露台与工作平台晒衣处

- 建议晾衣架应设在畅达的位置，最好设于室外。
- 为方便轮椅人士，建议晾衣架的高度不超过1200mm。如实际情况不能把晾衣架安装在1200mm以下，则须设置机械装置，把晾衣架拉低至适当的高度。

Location of a socket outlet
插座位置

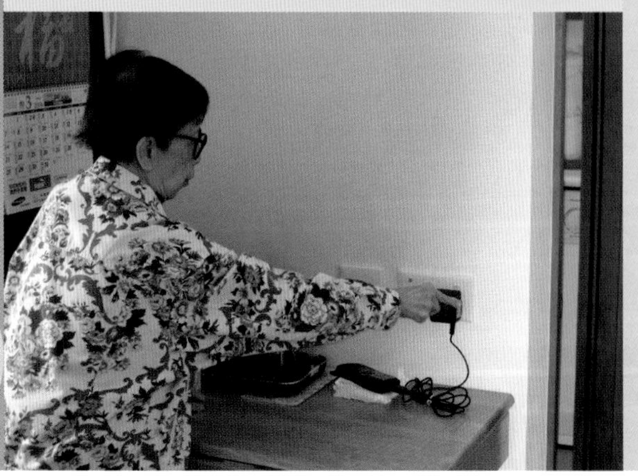

3

无障碍设计

Cha 章

Home Safety
家居安全

Home Safety

4.1 FIRE SAFETY

Fire safety is subject to complex and frequently updated legislation in statutory building regulation systems. Buildings shall at least comply with local regulations. Regulations may include specific, defined meanings to related terminology. The following is an introduction to related issues and should not be taken as indicative of statutory requirements.

4.1.1 Fire Prevention

Fire Safety Education and Training

- It is recommended that all residents, especially the elderly, children and people with disabilities, shall attend fire drill and fire safety briefings.

Storage Safety

- It is recommended that a lockable cabinet shall be provided to store volatile chemicals.
- It is recommended that lockable cabinets shall be provided in kitchens to store matches and lighters apart, especially from volatile chemicals.

Kitchen Safety

- Where provided, it is recommended that hooks for dishtowels shall be located not less than 1000mm away from stoves or cooktops.
- It is recommended that window curtains shall not be provided unless further than 1000mm away from stoves or cooktops.
- It is recommended that the functions of kitchen ventilation systems or rangehoods shall be checked to ensure mechanical function and to monitor build-up of airborne waste such as grease.

4.1

4.1 消防安全

消防安全受制于复杂和经常更新的法定建筑规范制度。建筑物最少应遵从本地规范，其可能包括有关术语的特定和界定意义。以下将推荐部分防火安全措施，不应视为法定要求。

4.1.1 防火措施

消防安全教育和训练

- 建议全体居民，尤其长者、儿童和残障人士，应该参与火警演习和消防安全讲座。

储物安全

- 建议应设置上锁的柜，用以贮存挥发性的化学品。
- 建议在厨房应设置上锁的柜，用以分开存放火柴及打火机，尤其是不要接近挥发性的化学品。

厨房安全

- 如厨房内设有抹布挂钩，建议应将其安装于离开厨房炉具至少1000mm的地方。
- 建议窗帘应离开厨房炉具至少1000mm。
- 建议应定期检查厨房通风系统或抽油烟机的排气功能，以确保其功能正常，清理由气体所产生的废物，例如油脂。

Dish towels are too close to the stove
抹布太接近炉具

Electrical Safety

- Where provided, it is recommended that electric heaters shall be installed more than 1000mm away from inflammables such as curtains. To allow heaters to be automatically switched off at night, it is recommended to installed time controllers.
- It is recommended that electrical outlets shall be located not less than 900mm away from water.
- Electrical outlets and extension cords should not be overloaded.
- Fixed electrical wiring should be concealed.
- Where exposed fixed electrical wiring already exists, it should be checked for damage regularly.
- Particularly where young children are active, it is recommended that unused socket outlets shall be capped.
- It is recommended that electrical applian-ces shall be unplugged and stored away from water when not in use.

电力安全

- 如设有电暖炉，建议应将其安装于离开窗帘等易燃物品最少1000mm的地方，并安装定时装置，在晚上自动关掉暖炉。
- 建议电源插座安装的位置，应离开水源最少900mm。
- 不可超载负荷电源插座和驳线。
- 固定的电线必须隐藏安装。
- 要定期检查外露的固定电线。
- 在小孩活动的地方，建议应将未使用的电源插座遮盖。
- 建议应拔除不使用的电器插头，贮藏在远离水源的地方。

The electrical outlet is too close to the sink and water tap
电源太接近洗涤盆及水龙头

The electrical wires are overloaded
电源超出负荷

4.1

4.1.2 Fire Escape Contingency Planning

It is recommended that every household shall develop a fire escape plan with full consideration for the needs of children, the elderly, the weak and people with disabilities.

As part of Fire Escape Contingency Planning, it is recommended that:

- Sketches indicating floor plans of flats and floor plans of the building shall be prepared, including all rooms with relevant features such as windows, doors, steps, stairways, ladders, lifts, smoke alarms, fire extinguishers, fire blankets, and furniture layout.
- At least two escape routes shall be identified to evacuate from a building, including provisions for the weak and people with disabilities.
- Heavy furniture or potentially dangerous furniture shall be located away from escape routes.
- Smoke alarms shall be installed in flats and communal areas.
- Smoke alarms shall be checked regularly to ensure functioning and user recognition.
- To ensure warning of residents with visual and hearing impairments, smoke alarms shall be selected with vibration, strobe light or other types of audible and visual signals.
- Where windows form part of an escape plan, windows shall be checked regularly to ensure ease of opening.
- A meeting place shall be designated, away from potential danger.
- Escape plan sketches shall be posted in all rooms and communal areas. Escape routes shall be indicated in contrasting colours.
- All residents shall participate in fire drills at least twice every year.
- All residents, particularly children, the elderly and people with disabilities, shall be informed of the location and functions of refuge areas where provided.

4.1.2 疏散计划

建议每个家庭都应制定一个疏散计划，并顾及老幼残弱的需要。

疏散计划应包含下列各项:

- 画出住宅单位和楼宇平面图，包括所有相关设施，如窗户、门户、梯级、楼梯、梯子、升降机、烟雾警报器、灭火筒、灭火毡及家具的位置。
- 至少应设定两条逃离楼宇的疏散路线，包括体弱者和残障人士适用的路线。
- 重型家具或有潜在危险的家具，应该远离逃生路线。
- 住宅单位和共用地方必须安装烟雾警报器。
- 必须定期检查烟雾警报器，确保功能正常，以及使用者能够辨识警报。
- 烟雾警报器应安装闪光灯、震动器或其他形式的听觉及视觉警号，确保视障及听觉障碍的居民可以收到警报。
- 如果窗户也为疏散设施的一部分，应定期检查窗户，确保能正常开启。
- 在远离有潜在危险的地点设定集合处。
- 所有房间和共用地方均须张贴疏散计划图。疏散路线应以对比色标示。
- 所有居民每年均须参加至少两次火警演习。
- 应通知所有居民，包括儿童、长者和残障人士有关避难区（如有的话）的位置和作用。

4

4.1.3 Barrier-Free Access of Escape

4.1

4.1.3.1 Refuge Areas

Where refuge areas are provided, accessibility is a priority. Refuge areas can be located in the following places:

(1) Protected Lobbies Adjacent to Staircases
(2) Protected Lobbies Adjacent to Fireman's lifts
(3) Stairway Smoke-Proof Enclosures
(4) Exterior Balconies Adjacent to Staircases
(5) Vented Stairway Landings

4.1.3 无障碍疏散设计

4.1.3.1 避难区

如果设置避难区，首要应考虑无障碍设计。避难区可设于下列各处:

(1) 靠近楼梯的消防前室
(2) 靠近电梯的消防前室
(3) 防烟区楼梯
(4) 靠近楼梯的室外露台
(5) 有自然通风设备的楼梯平台

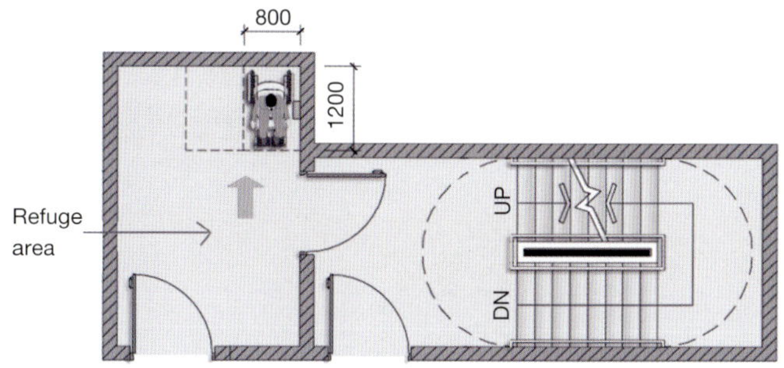

A stairway smoke-proof enclosure as a refuge area
有防烟围封的楼梯前室作为避难区

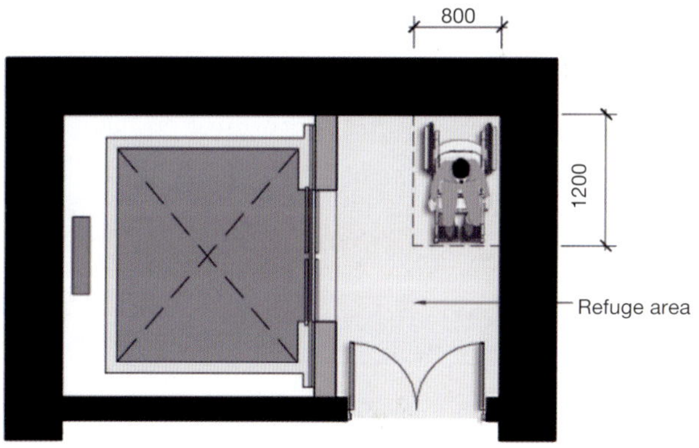

A protected lobby adjacent to the fireman's lift as a refuge area
靠近消防员升降机的前室作为避难区

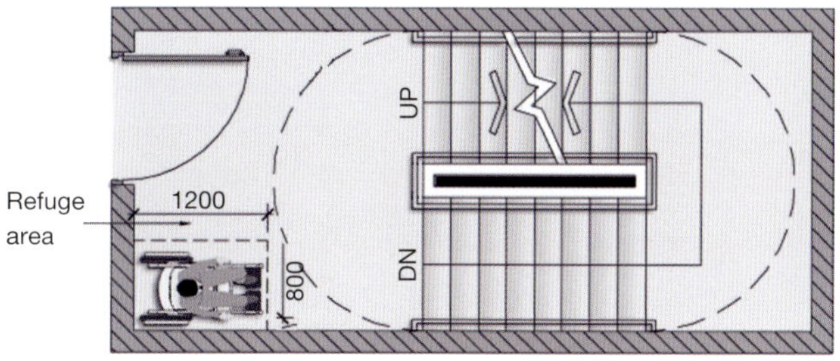

A vented stairway landing as a refuge area
有通风设备的楼梯上落处作为避难区

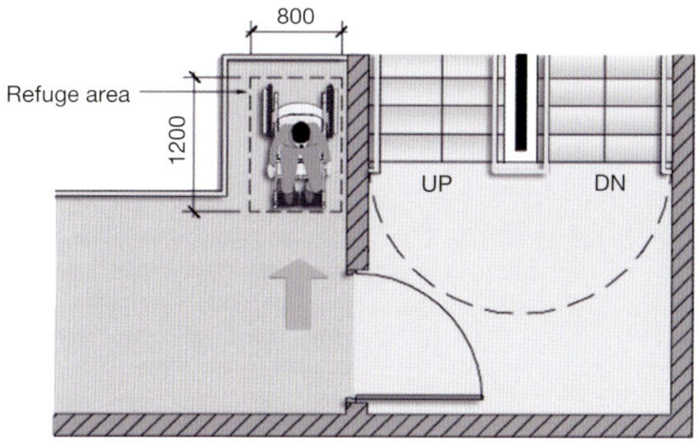

An exterior balcony adjacent to the staircase as a refuge area
靠近楼梯的外围露台作为避难区

4.1

Ideally, natural ventilation shall be provided to protected lobbies. It is recommended that design of refuge areas shall meet the following requirements:

Floor Surfaces for Refuge Areas

- To ensure recognition by people with visual impairments, it is recommended that tactile positional tiles shall be provided beside the entrance to a refuge area.
- Slip-resistant floor surfaces should be provided in refuge areas.

Fire Doors to Refuge Areas

- Refuge areas shall be enclosed by fire doors.

Wheelchair Parking Space in Refuge Areas

- At least one wheelchair parking space shall be provided in each refuge area.
- It is recommended that wheelchair parking spaces shall be indicated by distinctive floor materials or floor markings.
- It is recommended that the area of each wheelchair parking spaces shall be not less than 800mm x 1200mm.
- It is recommended that wheelchair parking spaces shall not block exit paths to staircases or fireman's lifts.

Fire Safety Equipment

- It is recommended that the refuge areas be provided with clear signage at the entrance to the refuge area and on escape routes. To enable recognitions by people with visual impairments, it is recommended that signs shall employ contrasting colours, and shall be duplicated by Braille and raised letter labelling in appropriate locations.
- To facilitate use by people in wheelchairs, it is recommended that communication equipment in refuge areas shall be provided close to wheelchair parking space and at a height between 750mm and 1100mm. It should enable calls to the Building Management Office and emergency services. To facilitate use by people with visual impairments, it is recommended that Braille and raised letter / character labelling shall be provided to all communication equipment.
- It is recommended that emergency call buttons be of a suitable type for use by people with disabilities and the elderly. To enable recognition by people with visual impairments, it is recommended that emergency call buttons shall employ contrasting colours, and shall be labelled in Braille and raised letters.

最理想的避难区应有自然的通风系统。建议避难区的设计应符合以下要求:

避难区的地面

- 建议在避难区入口的旁边，设置方位引路砖，以方便视障人士识别避难区入口方位。
- 避难区应铺设防滑地面。

避难区的防火门

- 避难区应设防火门。

避难区的轮椅停放处

- 每个避难区至少应设置一个轮椅停放处。
- 建议将轮椅停放处地面铺设显眼的物料或标示范围，以容易识别。
- 建议每个轮椅停放处的面积应不少于800mm x 1200mm。
- 建议轮椅停放处不应阻塞疏散至楼梯或消防升降机的通道。

消防设备

- 建议在避难区入口和逃生通道应设指示牌，明确标志避难区方位，以确保视障人士能够识别。建议这些标志应采用对比色，并于适当位置多次使用设有凸起盲文及字母的标志。
- 为了方便轮椅人士，建议在避难区靠近轮椅停放处的地方设置通信设备，高度应介 750 ~ 1100mm。这些通信设备可以联系至大厦管理处和紧急救援服务。为了方便视障人士，这些通讯设备应设置凸起盲文及字母。
- 建议应装置适合残障人士及长者使用的紧急救援按扭。为使视障人士能够识别，建议紧急救援按钮必须采用对比色，并设凸起盲文及字母。

4

4.1

4.1.3.2 防火门

防火门的设计应全面顾及残障人士和长者的需要，并建议符合下列要求:

- 建议防火门须适合残障人士使用，尤其轮椅人士。设计须符合第3章的无障碍门的设计要求。
- 建议所有疏散门的净宽不少于850mm。
- 建议疏散门应采用对比色，以方便视障人士。
- 建议设置发声信号，以指示疏散门方向，协助视障人士。
- 建议疏散门应设置观望板，建议板段底部不应离地面多于1000mm；其顶部不应离地面少于1500mm。
- 建议自掩门设置可调整的关门装置。

4.1.3.2 Fire Doors

Fire doors design should take into consideration the needs of people with disabilities and the elderly. It should meet the following requirements:

- It is recommended that fire doors shall be accessible to people with disabilities, especially people in wheelchairs. Design requirements for accessible doors are discussed in Chapter 3.
- It is recommended that the clear width of all escape doors shall not be less than 850mm.
- Particularly to facilitate recognition by people with visual impairments, it is recommended that escape doors shall be painted in contrasting colours.
- Particularly to facilitate recognition by people with visual impairments, it is recommended that an audible signal shall be provided to indicate the location of escape doors.
- It is recommended that vision panels shall be provided to escape doors. It is recommended that the bottom edge of the vision panel shall be at a height of not more than 1000mm and the top edge at a height of not less than 1500mm above floor level.
- Where there is a self-closing door, it is recommended to provide an adjustable closing mechanism.

4.1.3.3 消防员升降机

- 多于一层的楼宇，建议消防升降机应可到达所有楼层，包括避难层。
- 建议消防升降机应附设消防楼梯。
- 建议消防升降机须为无障碍升降机。无障碍升降机的设计参见3.8节。
- 建议大厦应有主要及辅助电源供应以防火灾将主要电源供应中断。
- 建议消防升降机应设置通信器材，以便传唤大厦管理处和紧急救援服务。建议所有通信器材应设置凸起盲文及字母/文字标志，以方便视障人士使用。
- 建议消防升降机应设置自行返回地面的按钮。
- 消防升降机及其附设楼梯须用耐火材料围护封闭。

4.1.3.3 Fireman's Lifts

- It is recommended that for buildings of more than one floor, all floors, including refuge floors shall be accessed by fireman's lifts.
- It is recommended that fireman's lifts shall be combined with a fire-rated staircase.
- It is recommended that fireman's lifts shall be designed to be accessible lifts. Design of accessible lifts is discussed in chapter 3.8.
- In case a fire breaks in the main supply, it is recommended that fireman's lifts shall be provided with duplicate power supply.
- It is recommended that fireman's lifts shall be provided with communication equipment to enable calls to the building management offices and emergency services. To facilitate use by people with visual impairments, it is recommended that Braille and raised letter / character labelling shall be provided to all communication equipment.
- It is recommended that fireman's lifts shall be provided with a push button for automatic return to ground floor.
- Fireman's lifts and associated staircases shall be protected by a fire-rated enclosure.

Home Safety

Visual Alarm
视觉警报

4.2

4.2 ALARM AND EMERGENCY PROVISIONS

4.2.1 Audible Alarm and Public Announcement Systems

- It is recommended that combined alarm systems shall be provided to alert all residents in case of emergency.
- It is recommended that public announcement systems shall be provided at corridors, staircases and protected lobbies. To minimise panic, emergency messages should be calm in tone.

4.2.2 Visual Alarm Systems

In order to alert people with hearing impairments, it is recommended that:

- Visual alarms shall be provided in addition to audible alarms.
- Tests and fire drills shall be undertaken to ensure visual alarm systems are adequately bright.
- Alarm systems with vibration shall be provided.

4.2.3 Fire Blankets

- It is recommended that fire blankets shall be provided in kitchens.
- To facilitate use by children and people in wheelchairs, it is recommended that fire blankets shall be located with their pull tapes at a height of not more than 1200mm above floor level.
- It is recommended that fire blankets shall be labelled in Braille and raised letters / characters no less than 25mm high.

4.2.4 Fire Extinguishers

- It is recommended that fire extinguishers shall be provided in kitchens, communal corridors, and outside protected lobbies or staircases.
- To facilitate use by children and people in wheelchairs, it is recommended that fire extinguishers shall be hung with their handles not higher than 1200mm above floor level.
- It is recommended that fire drills shall offer demonstration of fire extinguisher use.
- To facilitate use by the weak, the elderly and children, it is recommended that small, portable fire extinguishers shall be provided.

4.2 警报及应急装备

4.2.1 发声警报及公众广播系统

- 建议应使用综合式的警报系统，在紧急时警告所有居民。
- 建议应在各走廊、楼梯及消防前室大堂装置公众广播系统。紧急广播信息须语调平静，以减低恐慌。

4.2.2 视觉警报系统

建议设置以下设施，以方便听觉障碍人士:

- 在发声警报以外，另设置视觉警报设备。
- 要进行测试和演习，确保视觉警报系统的光度充足。
- 应设有震动装置的警报设备。

4.2.3 灭火毡

- 建议在厨房设置灭火毡。
- 建议灭火毡拉条高度不应离地面超过1200mm，方便轮椅人士及儿童使用。
- 建议灭火毡的标签应设有凸起盲文及字母/文字，其字体高度不应少于25mm。

4.2.4 灭火筒

- 建议厨房、公共走廊、消防前室外和楼梯均应设置灭火筒。
- 建议灭火筒把手高度离地面不应超过1200mm，以方便轮椅人士及儿童使用。
- 建议应该进行火警演习，讲解灭火筒的使用方法。
- 建议提供小型、易携的灭火筒，方便体弱者、长者和儿童使用。

4

4.3 LIGHTING AND SIGNAGE

4.3

4.3.1 Lighting

- It is recommended that communal area, such as corridors and lift lobbies, shall be provided with artificial lighting to give illumination of not less than 120 lx at floor level. In order to save energy, it is recommended that where natural light is available in daytime, or usage is intermittent, sensors and timers shall be provided to control lighting.
- It is recommended that an illumination level of not less than 120 lx shall be provided on the face of signs.
- To facilitate use by people with visual impairments, it is recommended that higher illumination shall be provided at locations requiring activities, such as flat entrances and refuse chutes.

4.3 照明设备和指示牌

4.3.1 照明设备

- 建议在走廊和升降机大堂等共用地方设置人工照明系统，地面照明度不得少于120lx。为节省能源，建议在可使用自然光线的日间，或在间歇使用期间，应使用感应器及定时器控制人工照明时间。
- 建议在指示牌表面的照明度不得少于120lx。
- 建议在活动频密的地方如住宅单位入口及垃圾槽，设置较光亮的照明，以方便视障人士使用。

Bad practice- an internal corridor with insufficient lighting
不良做法- 无充足照明的室内走道

4.3

4.3.2 Signage

General Design Considerations

- To facilitate understanding by people with visual impairments, it is recommended that signs shall be supplemented by labelling with Braille and raised letters / characters of 60mm height.
- It is recommended that signs shall be written in both English and Chinese.
- It is recommended that functional signs shall give clear instructions or information, with easily understood graphics.

Locations

- The location and design of signs is an independent design discipline. Certain notices, labels and directional signs may also be subject to statutory requirements. It is recommended that simple labels shall be provided between 800mm and 1500mm above floor level.
- When refuge floors are provided, it is recommended that their location shall be indicated in both English and Chinese at the main entrance of the building.
- It is recommended that signs at flat entrances indicating flat numbers shall be provided at a maximum height of 2000mm.
- It is recommended that maps and information panels at building entrances and along roads shall be provided at a height between 900mm and 1800mm.
- Overhanging signs should allow a minimum overhead clearance of 2000mm.
- It is recommended that signs shall be provided at the lift lobbies and stair landings to inform the floor number.
- To prevent reflection, it is recommended that signs shall not be placed behind glass.

Shape of Signboards

- It is recommended that:
  (a) Information signboards shall be rectangular.
  (b) Warning signboards shall be triangular.
  (c) Interdictory signboards shall be circular.

4.3.2 指示牌

一般设计考虑

- 建议指示牌设置凸起盲文及字母/文字，以方便视障人士使用，字体高度宜不少于60mm。
- 建议指示牌应用中英文书写。
- 建议指示牌的资料和指示要清晰，并有容易理解的图像表示。

位置

- 指示牌的设计及摆设位置是一门专门的学问。某些通告、标志和指示牌也须符合法定要求。建议简单的标记应该离地面800～1500mm。
- 楼宇如设有避难层，建议应在楼宇主要入口处提供中英文指示，显示避难层的位置。
- 建议住宅单位门口显示门牌号数的指示牌，应位于离地面不应超过2000mm的地方。
- 建议应在楼宇的入口和沿途，提供地图及讯息指示牌，其高度应介乎900～1800mm。
- 悬挂指示牌下净空高度至少应有2000mm。
- 建议应在升降机大堂及楼梯平台提供指示牌，以显示层数。
- 建议所有指示牌不应放在玻璃后面，避免反光。

指示牌形状

- 建议如下:
  (a) 讯息指示牌宜为长方形。
  (b) 警告指示牌宜为三角形。
  (c) 禁止指示牌宜为圆形。

Signs beside the accessible lift door
在方便易达升降机门旁的指示牌

Signs at the lift lobby
升降机大堂指示牌

Signage for M&E room
电机房指示牌

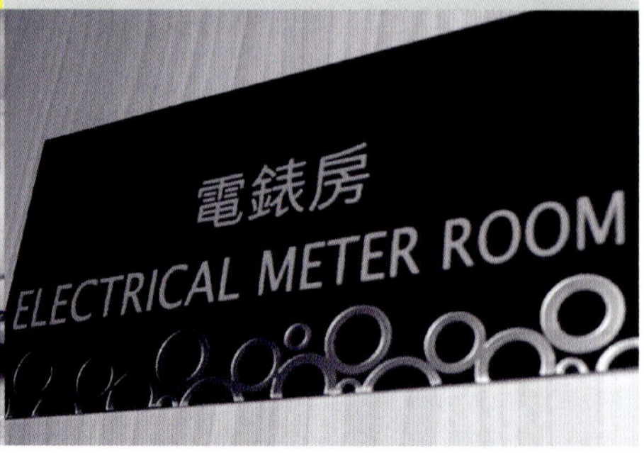

4

字体

以下提供一些字体大小的设计指南:

位置	字体大小
门口	60mm
走廊	110mm
室外	200mm

- 建议文字要用粗体，字高不得少于60mm，以方便视障人士使用。
- 建议字体大小要与阅读距离成正比。
- 建议指示牌采用对比色，附以凸起盲文和文字/字母对所传递信息做解释，以方便视障人士使用。
- 建议凸起文盲和中英文字体使用至少应凸起1mm，以方便视障人士使用。

火警疏散指示牌

- 火警疏散指示牌须符合法定要求。
- 此外，建议指示牌应能清楚标示往疏散楼梯、避难区及消防升降机的方向，并应能顾及老弱和残障人士的需要。

Size of letters for signage

指示牌字体大小

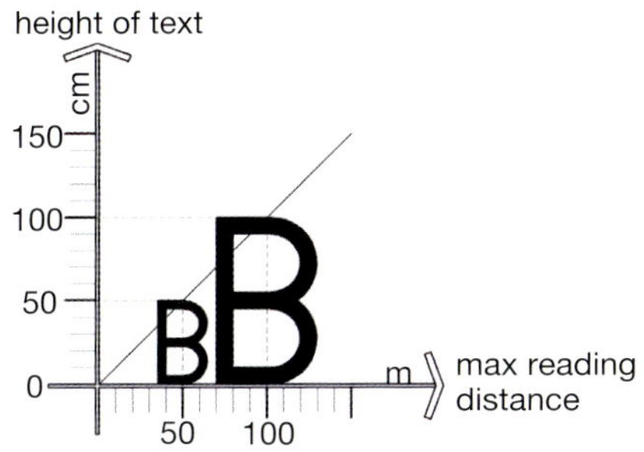

Fonts

The following is offered as guidance to the font size of signs:

Location	Font size
Doors	60mm
Corridors	110mm
External use	200mm

- To facilitate understanding by people with visual impairments, it is recommended that letters and characters shall be in bold type not less than 60mm high.
- It is recommended that the size of letters shall be in proportion to the reading distance.
- To facilitate understanding by people with visual impairments, it is recommended that signs shall be in contrasting colours, with supplementary labelling in Braille and raised letters / characters.
- To facilitate use by people with visual impairments, it is desirable that the Braille and letters / characters shall be raised at least 1mm from the background.

Fire Escape Signs

- Fire escape signs are subject to statutory requirements.
- In addition, it is recommended that signs shall clearly indicate the directions of fire escape staircases, refuge areas and fireman's lifts, with considerations for the weak, the elderly, children and people with disabilities.

Bad practice- the contrasting colour does not emphasize the information

不良做法—所使用对比色无法强调所提供信息

Home Safety

4.4 PREVENTION OF INJURY

4.4.1 Fall Prevention

Falls are a major cause of home injury for both children and the elderly. In a local study, 18% of men and women aged 70 and above reported a fall during the past 12 months. Among these, about 4% of the men and 8% of the women sustained fractures. Apart from physical injury, falls can have effects such as dehydration, pressure sore, pneumonia and rhabdomyolysis. Environmental hazards are one of the main causes of falls. According to another local study, slipping, tripping over or kicking objects were major reasons. In order to reduce the risk of falling, it is recommended to address accessibility design issues discussed in Chapter 3 as well as the following:

Window Safety

Children are at high risk from window-related falls. To prevent window falls:

- It is recommended that guards shall be installed on all windows to prevent children from falling out of windows.
- Where window guards are not provided and children have access, it is recommended that furniture shall be located away from windows to prevent children from climbing up and reaching windows.
- It is recommended that windows shall be checked regularly by professional technicians at least every 2-3 years to ensure ease of opening and secure closure.
- Window cords can pose a hazard to infants and young children. It is recommended that cords shall be secured out of the reach of children.

Bathroom Safety

- It is recommended that slip-resistant mats or adhesive safety strips shall be provided in bathtubs and showers.
- It is recommended that nightlights shall be provided to corridors and bathrooms.
- It is recommended that an upstand shall be provided at the edge of washbasin counters to prevent water from dripping on the floor.
- It is recommended that rubber mats shall be placed in front of sinks to prevent slipping.

4.4 预防创伤

4.4.1 防止跌伤

长者和儿童遇到的家居意外主要是跌伤。根据一个本地研究显示，有18%年龄超过70岁的男士和女士，在过去12个月内曾经跌伤一次。其中约有4%的男士和8%的女士骨折。跌伤除了会造成身体伤害，还会带来以下的结果，例如脱水、受压疼痛、肺炎及横纹肌溶解。环境的危险是跌伤的主要原因之一。根据另一个本地研究显示，滑倒或绊倒物件是跌伤主要的原因。为了减低跌伤的危险，本指南除了在第3章讨论无障碍设计外，还建议注意下列事宜:

窗户安全

儿童是从窗户跌下的高危人士。要预防从窗户跌下，可参考下列做法:

- 建议在所有窗户装置窗防护栅栏，防止儿童从窗户跌下。
- 在没有装置窗防护栅栏及儿童可及的窗户，建议应将家具搬离窗户，使儿童不能攀越。
- 建议每隔两三年为窗户作专业的技术检查，以确保容易开启及能紧闭。
- 窗帘绳对婴儿和小孩可能造成危险，建议应该绑好绳索，避免儿童触及。

浴室安全

- 建议在浴缸和淋浴间装置防滑垫或安全粘贴条。
- 建议在走廊和浴室设置夜灯。
- 建议在洗面盆柜面的边缘安装挡台，防止有水流到地面。
- 建议在洗面盆前面放置塑胶地垫，防止滑倒。

家居安全

Kitchen Safety

- It is recommended that lighting be provided to give illumination of not less than 120 lx at worktop level.
- It is recommended to round the corners of kitchen counter tops.
- It is recommended that upstands be provided at the edge of sinks to prevent water dripping on the floor.

Furniture

- In order to facilitate use by the elderly and people in wheelchairs, it is recommended that furniture be arranged to provide sufficient walking space or manoeuvring space.
- In order to facilitate use by the elderly, it is recommended that chairs and sofas should not be too low or close to the floor. Preferably, chairs and sofas should have firm base supports with a height of not less than 450mm from the surface of the seat.
- To avoid failure, it is recommended check the backs and arms of chairs regularly.
- Where carpet is used, it is recommended that it should be tacked down to avoid slippage.

厨房安全

- 建议在工作台提供不少于120lx的照明。
- 建议将厨房的柜台角改为圆角。
- 建议在洗涤盆边缘安装挡台，防止有水流到地面。

家具

- 建议家具的摆设应预留充足的步行或轮椅操作空间，方便长者和轮椅人士使用。
- 建议椅子和沙发不应太矮及过于贴近地面，以方便长者使用。椅子和沙发应有稳固的背部支撑，其高度由座位计算应不少于450mm。
- 建议应定期检查椅背和椅柄，防止损坏。
- 如果使用地毯，建议应将之粘牢。

Window guard
窗花

Home Safety

4.4

4.4.2 Prevention of Poisoning

To prevent from poisoning, it is recommended that:

- Gas valves shall be clearly marked with on and off positions.
- Child safety locks shall be provided on cabinets where chemicals such as medical, personal care and cleaning products are stored.
- Carbon monoxide alarms shall be provided in kitchens, bathrooms and bedrooms to prevent carbon monoxide poisoning.
- All habitable rooms shall have natural ventilation, including kitchens and bathrooms.

4.4.2 预防中毒

建议实施以下措施，以防止中毒:

- 应该清楚标示煤气管道上的开关位置。
- 摆放药物、个人护理及清洁用品的橱柜应设有儿童安全保护锁，防止儿童拿取。
- 厨房、浴室和卧室应装置一氧化碳警报器，防止发生一氧化碳中毒。
- 所有居住房间包括厨房和浴室，应有自然通风。

4.4.3 Emergency Management

- It is recommended that emergency call buttons or pull cords shall be provided in kitchens and bathrooms for the elderly.
- To deal with problems experienced by the elderly, it is recommended that individuals join local emergency call system, such as the Personal Emergency Link [PE Link] system. Users carry a remote controlled device incorporated into a wrist-watch or necklace. In case of emergency users can press the control button to notify a call centre. The location of users can be detected for prompt assistance.
- To facilitate access in an emergency, it is recommended to provide emergency release locks where possible.
- To facilitate elderly's use, it is recommended that telephones with large numeric keypads and enhanced audio systems shall be provided in an accessible location.
- To deal with problems experienced by the elderly, it is recommended that emergency telephone numbers shall be posted boldly in prominent locations.

4.4.3 应急措施

- 建议在厨房和浴室为长者设置紧急召援把手或拉绳。
- 建议使用紧急呼援服务，如一线通呼援服务中心的服务，以助长者解决问题。使用者身上可携带一种手表或颈链般的紧急召援的装置，遇到紧急事故时可按钮通知召援中心，中心可以侦察到使用者的位置，及时提供救援。
- 若情况可行，建议装置紧急事故时可自动或由他人开启的锁，方便有关人士在紧急事故时进入该处。
- 建议在容易拿取的位置安装设有大型数字和有助听系统的电话，以方便长者在紧急情况时使用。
- 建议在显眼位置张贴大字体的紧急事故电话号码，以供长者在紧急情况时使用。

4

Emergency call button
紧急呼援服务

A telephone with a large numeric keypad
大型数字键盘的电话

家居安全

章

Cha

pter

5

Environmental Factors
环境因素

Environmental Factors

5

5.1 HEALTH-PROMOTION

In addition to traditional and passive public health considerations, Universal Design seeks to engage with positive issues of wellness. In housing one aspect of that is to create opportunities for all residents to participate conveniently in physical exercise.

5.1.1 Staircases for Health-Promotion

Staircases have potential to provide more than vertical circulation and means of escape. With appropriate design, staircases can act as social spaces and encourage residents to use them for exercise. To provide user-friendly staircases, it is recommended that:

- Staircases shall be cleaned regularly.
- Rubbish bins shall not be located in staircases. This is easier to achieve when convenient designated disposal locations are available.
- Staircases shall be designated to provide natural light and ventilation. Where natural light is not available, it is recommended that illumination of not less than 120 lux shall be provided. To save energy, it is recommended that sensors and timers shall be provided to control lighting. Where natural ventilation is not available, quiet and efficient mechanical ventilation shall be provided.

5.1.2 Laundry for Health-Promotion

It is recommended that natural drying facilities shall be provided. To avoid nuisance in high-rise and other intensively managed developments, it is important that natural drying locations shall be designated at the design stage. The following options can be considered.

- Drying Balconies.
- Drying space on building roofs.
- Naturally-ventilated utility spaces inside flats.
- It is recommended that fixed drying racks shall be located safely and conveniently, with consideration for the elderly and people with disabilities. Soffit-hung drying racks shall be easily lowered. Window or balcony hung racks shall be accessible without leaning out. Construction shall be durable and sturdy.
- It is suggested that louvers can be utilized as part of building design, to screen hanging laundry from public view.
- It is recommended that freestanding drying racks shall have a height of between 900mm and 1200mm, foldable to save space.

5.1 促进健康的设计要素

通用设计住宅除了考虑传统和间接促进公共卫生的设计元素外，还应注意提供促进健康的直接设计元素。其中一方面就是为所有居民提供多做运动的机会。

5.1.1 促进健康的楼梯

楼梯不单供人上落或疏散之用，如果设计合宜，楼梯还可以用作提供社交空间或让居民锻练身体。一条方便居民使用的楼梯，建议以下做法:

- 定期清理楼梯。
- 楼梯不应放置废物箱。通过设置方便易达的垃圾存放处，可简单达到此目的。
- 楼梯应使用自然采光和自然通风。如果无法提供自然采光，则建议设置照明度达120lx的照明系统。建议使用感应器和定时器控制照明时间以达到节能目的。如果无法提供自然通风，则应装置噪声低和高效能的机械通风设施。

5.1.2 促进健康的晒衣处

建议装置自然通风晾晒的设施。为免对高楼和对密度高的建筑物造成干扰，应在设计阶段考虑自然通风晾晒处的设置，可考虑以下相应措施:

- 设置晾晒露台。
- 楼宇天台有晾晒空间。
- 住宅单位内有自然通风的工作平台。
- 建议装置固定的晾晒架于容易到达和安全的位置上，以方便长者和残障人士使用。活动的晾晒架必须容易取下。置于窗口或露台的晾衣架必须容易使用，使用者无需俯身向外。结构必须稳固耐用。
- 建议在楼宇立面相应位置安装百叶板遮掩晾晒的衣物，以减少晾晒衣物对楼宇外观的影响。
- 建议立式晾晒架高度应介乎900~1200mm，并可折叠，腾出更多空间。

环境因素

- Where it is difficult to provide naturally-ventilated spaces, built-in ceiling mounted dehumidifiers can be provided as a substitute in toilets. This shortens drying time and also keeps indoor humidity at a suitable level [See Chapter 5.3].

- 如不能提供自然通风的晾晒空间，可在厕所顶棚设置内置抽湿机代替。这不但能减少晾晒时间，还能使室内湿度保持适中[参照第5.3节]。

Drying racks outside the building cause nuisance and dripping water to below
露天晾衣架不但造成干扰，亦可能向下层滴水

5.1

5.1.3 Landscape Garden and Open Areas for Health-Promotion

Landscape garden and open spaces as part of residential development can encourage wellness related activities:

Planting for Health-Promotion

Planting can be far more providing than just visual amenity. It can be an element for health-promotion. With consideration for health-promotion, it is recommended that:

- Plants with thorns shall be avoided, except as a security measure.
- Plants shall be non-toxic.
- Plants shall have a variety of colours and scents to stimulate interest.

Design for Health-Promotion

- To encourage regular exercises, it is suggested that physical exercise settings, such as exercise stations and foot massage paths, can be provided in open areas.
- For exercise and personal expression, it is suggested that self-planting corners can be made available to entertain residents.
- To provide opportunities for social activities, it is suggested that features are able to act as a social focus, such as chess tables, and can be provided in open areas.

Ancillary Facilities in Neighbouring Open Areas

- It is recommended that seating and social areas shall be provided, with opportunities for sun and shade.
- It is recommended that adequate rubbish bins shall be provided.
- It is recommended that vending machines shall be provided.
- It is recommended that accessible public toilets shall be provided.
- It is recommended that lightings shall be provided at a sufficient illumination level, ideally 75 ~ 120 lx.

A self-planting corner in neighbouring open area
附近开放空间的自行种植地

Rubbish bins in neighbouring open area
附近空地的废物箱

5

5.1.3 促进健康的园林花园及开放空间

园林花园和开放空间作为住宅建设的一部分，可以促进优质生活：

促进健康的植物

植物不单可增进视觉享受，更是促进健康的元素。促进健康的建议如下：

- 除非作为保安措施，应避免栽种有刺植物。
- 栽种无毒植物。
- 栽种不同颜色和香味的植物，以刺激官感。

促进健康的设计

- 建议在露天场地设置健身器和足部按摩径等运动设施，以鼓励居民坚持锻练身体。
- 建议设置可供住户自行栽种植物的地方，以增添居民的生活乐趣和增加居民运动的机会。
- 建议在露天开放空间放置棋桌等社交设施，以增加社交活动的机会。

临近露天开放空间的附属设施

- 建议设置座椅和社交区，使其有阳光照射，并附设雨棚等遮荫设施。
- 建议摆放足够的废物箱。
- 建议设置自动售卖机。
- 建议设置无障碍公共厕所。
- 建议提供充足的街灯照明，理想照明度为75～120lx。

Exercise stations in neighbouring open area
附近空地的健身设施

A massage path in an neighbouring open area
住宅附近开放空间的按摩径

5.1

5.1.4 Indoor Communal Spaces for Health-Promotion

In order to facilitate family activities and encourage neighbour's interaction, it is recommended to provide:

- Quality indoor communal space within residential blocks or within clubhouses;
- Comfortable seating and social corners in indoor communal spaces;
- Physical exercise equipment in indoor communal spaces; and
- Facilities to encourage family and social activities.

5.1.4 促进健康的室内共用空间

建议提供下列设施，以促进家庭和邻舍的关系：

- 在楼宇或会所内，设置优质共用空间。
- 在室内共用空间设置舒适座椅和交谊区
- 在室内共用空间，提供运动设施。
- 提供鼓励家庭和社交活动的设施。

5.1.5 Facilities for Pets for Health-Promotion

Pets could provide emotional support to people and facilitate psychological health. Where residential developments allow pets, it is recommended that acceptable parameters shall clearly be defined for issues such as species, likelihood of injury, noise, odour, disease and waste. Within these acceptable parameters, it is recommended that facilities shall be provided for the convenience of pet owners and other residents:

- Pets may generate a wide variety of waste, some of it with potential to communicate disease. Thus, acceptable means of disposal need to be identified and notified to pet owners.
- The presence of some pets may be acceptable for periods in communal areas. To facilitate use by the weak or people with disabilities, it is suggested that leash points can be provided for dogs in locations such as communal seating areas.
- It is suggested that hanging points can be provided for birdcages in communal open areas.
- It is suggested that service lifts can be used by residents with pets.

5.1.5 促进健康的宠物设施

宠物可提供情感支持，并促进心理健康。如在居住环境饲养宠物，建议须考虑一些认可条件，如宠物品种、宠物会否受伤害、宠物的声音、气味、疾病和排泄物等。在符合有关认可条件下，建议应向宠物主人和居民提供方便的设施。

- 宠物会排出各种排泄物，有些可能会传播疾病。应该指示宠物主人处理排泄物的认可方法。
- 在某些时段和在某些共用地方，携带宠物是可以接受的。建议在共用地方设置杆柱围栏，以捆绑狗只的链带，以方便体弱和残障人士。
- 建议在露天的公共地方设置栅架，供悬挂鸟笼。
- 建议携带宠物的居民，应使用专用或服务升降机。

5.2

5.2 INDOOR AIR QUALITY

Indoor air quality (IAQ) has an immediate impact on the health and wellbeing of all building users, through a combination of factors such as temperature, humidity, flow rate and airborne materials. All of these are mediated by buildings, though how and to what extent may vary considerably. In the Hong Kong urban area, airborne particulates from diesel exhausts have been a significant pollutant which can be largely avoided in sealed buildings with mechanical ventilation. In other circumstances, buildings can contribute to airborne pollution too. Notably, many building materials, furnishing, and household cleaning and maintenance materials continuously leak small quantities of chemicals into the air.

5.2 室内空气质量

室内的空气质量会直接影响所有居民的健康，这些影响来自多种结合的因素，如温度、湿度、气流和空气传播的物质等。这些影响因素都来自建筑物本身，纵然影响的程度和方式各有不同。在香港市区，空气所传播来自油渣废气的微粒是最主要的污染物，但在密封的设有机械通风系统的建筑物内，这种污染可以减少。但在另一些情况下，建筑物本身也可造成空气传播的污染，最明显的例子是建筑物料、装修物料和家居清洁用品及维修物料等，都会不断向空气释出少量的化学物质。

5

环境因素

室内的空气污染物包括空气浮尘、悬浮生物微粒、二氧化碳、一氧化碳、二手烟、甲醛、臭氧、挥发性的有机化合物和氡。要去除污染物的源头，建议应用下列策略：

5.2

5.2.1 去除污染源

- 油漆是由颜料、媒剂和其他化学品合成的，这些物品中的任何一种都可能包含毒素。彻底解决的方法就是去除有毒油漆，因为在一些情况下，油漆的主要功用只是为了装饰，但是有些时候，使用油漆也是为了延长所用物料的寿命。在这种情况下，耐用与毒素二者的优劣便只能选择其一了。有些时候，在近海的环境或在历史建筑物内不容易到达的地方，必须使用较耐用的油漆，而毒素须减至最低水平。为了尽量减少挥发性的有机溶剂释入空气中，我们建议在大多数的室内设施，避免使用矿物性的油性油漆。很多时候，是可以用传统的高性能水性油漆取而代之的。
- 所有塑料和涂上塑料的物质都含有挥发性的有机溶剂，特别是包含人造纤维的地毯和纺织品，包括衣物。涂上防渗漏剂的木块或石块不一定会较好。因此必须在每个特定环境下作出取舍。可是一般来说，我们建议在室内应尽量避免使用塑胶涂料或人造纤维的物质。

Indoor air pollutants include airborne dust, bioaerosols, carbon dioxide (CO_2), carbon monoxide (CO), environmental tobacco smoke, formaldehyde, ozone (O_3), volatile organic compounds (VOC's) and radon. To remove the source of pollutants, the following strategies are recommended:

5.2.1 Removal of Pollutants' Sources

- Paints are a combination of pigments, media and other chemicals, any of which can be toxic. Thus, elimination of toxic painted finishes is a fundamental approach to this issue in circumstances where paint is primarily decorative. Where however the function of paints is to prolong the useful life of materials to which they are applied, a judgement sometimes has to be made between the durability and toxicity of the coating. It might be necessary to use a more durable paint in a marine environment or an inaccessible location in an historic building, whereas toxicity would undoubtedly be the priority at low level. In order to minimise the emission of volatile organic solvent, it is suggested that mineral oil-based paints can be avoided in most internal applications. A variety of traditional and high performance modern water-based paints is available to deal with most situations.
- Volatile organic solvents are present in all plastic and plastic coated materials, particularly carpets and textiles containing man-made fibres, including clothing of users. Timber or stone are not necessarily friendlier if treated with sealants. Judgement has to be exercised in each circumstance to determine a viable natural alternative. In general however, it is suggested that plastic and plastic coated materials and man-made fibres can be avoided in most internal applications.

- Similar concerns apply to cleaning materials. Judgement has to be exercised in each circumstance. An unfinished natural material is not necessarily friendlier if it has to be bleached regularly with powerful chemical throughout its working life.
- Airborne dust may be unavoidable where natural ventilation is used, especially where there is heavy traffic. Dust is also generated by the natural dispersal of human skin. Where low maintenance or high standards of hygiene are required, it is recommended that the building interior shall be designed with smooth surfaces and no unnecessary projections.
- Where low maintenance or high standards of hygiene are required, it is recommended that
  (a) Furniture shall be designed with smooth surfaces and no unnecessary projections.
  (b) Fabric hangings shall be omitted or cleaned regularly.
  (c) Smooth-surface flooring materials shall be used, such as wood and linoleum. Where carpet is preferred, it shall be vacuumed daily.
  (d) Decorative elements shall be reduced to a minimum.
  (e) Kitchen worktops shall be designed with smooth, and continuous surfaces.
  (f) Painted finishes on walls shall be washable.
  (g) Duct cleaning devices shall be used regularly to maintain mechanical ventilation systems.

- 同样的顾虑也存在于清洁用品的使用中。在每种情况下要作出不同的判断，一件未有装饰面的天然物料不一定较好，比如这物料必须定时使用强力的化学品加以漂白。
- 在自然通风和交通频繁的环境下，空气传播尘垢似是无可避免的。尘垢其实也会由人的皮肤自然产生。如要尽量减少维修，又要保持高度卫生，建议室内的设计要尽量平滑，减少不必要的凸出物。
- 如要尽量减少维修，又要保持高度卫生，建议：
  (a) 家具设计要表面平滑，尽量减少凸出物。
  (b) 避免使用纤维悬挂物或定期清洁。
  (c) 尽量使用表面平滑的地板，如木块和漆地胶等。如要铺设地毯，则须每天吸尘清洁。
  (d) 饰物应减至最少。
  (e) 厨房橱柜的表面应平滑及连续。
  (f) 墙壁上的漆油应可洗刷。
  (g) 要安装风管的清洗装置，定期清理通风系统。

5.2.2 Ventilation

- It is recommended that natural ventilation shall be provided in all bedrooms, living / dining rooms and kitchens.
- It is recommended that natural or mechanical ventilation shall be provided in toilets.
- It is recommended that cross ventilation shall be provided.

5.2.2 通风

- 建议所有的卧室、客/饭厅和厨房都使用自然通风。
- 建议厕所应使用自然通风，或设置机动通风系统。
- 建议利用对流通风。

5.3 THERMAL ENVIRONMENT

- It is recommended that the temperature of all exposed surfaces and water supplies shall be less than 43°C.
- Where there is an imbalance of radiant heat resource, it is recommended that such imbalance shall not be more than a range of 10°C.
- Where floor heating systems are used, it is recommended that floor temperatures shall not exceed 29°C and shall be not more than 5°C higher than the ceiling temperature.

5.3 热环境

- 建议所有外露表面及水源的温度，不应超过43℃。
- 遇有辐射热源温度不平衡，温差不应大于10℃。
- 如果使用地板供热系统，建议地板温度应保持在29℃以下；且地面的温度不应较顶棚的温度高出5℃。

• 建议浴室应设置合适的暖气设备，以方便长者和体弱人士。
• 建议室内最佳温度应保持在22℃～24℃。
• 建议室内湿度应保持在70%以下。可使用抽湿机，以减低湿度。

5.4 视觉环境

有效的照明能改善生活环境的质量，特别是在晚上:

• 建议背景光线应有足够照明度并且光照均匀。
• 在市区，周遭楼宇的灯光或街灯可对居民造成干扰，特别是来自一些闪动性的光源，如霓虹灯或大型光管招牌。如有这类干扰发生，建议使用可调校的窗帘。
• 为方便视障人士，建议在大门入口、升降机门、扶手、指示牌和电源开关等位置使用对比色区分。「对比色」是指使用在颜色轮上对立的颜色，例如红与绿和蓝与黄都是对比色的搭配。如以影印或别的方法把颜色影像转为灰度，便可以看到颜色对比度是否足够。

• To facilitate use by the elderly and the weak, it is recommended that appropriate heating devices shall be provided in bathrooms.
• It is recommended that indoor temperature shall be maintained within a range from 22°C to 24°C.
• It is recommended that indoor humidity shall not exceed 70%. Dehumidifier can be used to reduce humidity.

5.4

5.4 VISUAL ENVIRONMENT

Efficient lighting is necessary to the quality of living environment, particularly at night:

• It is recommended that even background lighting shall be provided at sufficient illumination levels.
• In urban area, lighting from neighbouring buildings or streetlights can be a source of nuisance to residents, particularly if the source is intermittent, as with flashing signs or giant video screens. It is recommended to provide adjustable window screening where such avoiding of nuisance is envisaged.
• To facilitate use by people with visual impairments, it is recommended that access features, such as entrance doors, lift doors, handrails, signboards and light switches shall be provided in contrasting colours. "Contrasting colours" refer to opposite colours on a colour wheel. For example, green and red or blue and yellow, are contrasting. By photocopying or other methods to convert the images to a grey scale, it can be judged if the colours of the images are at a sufficient contrasting level.

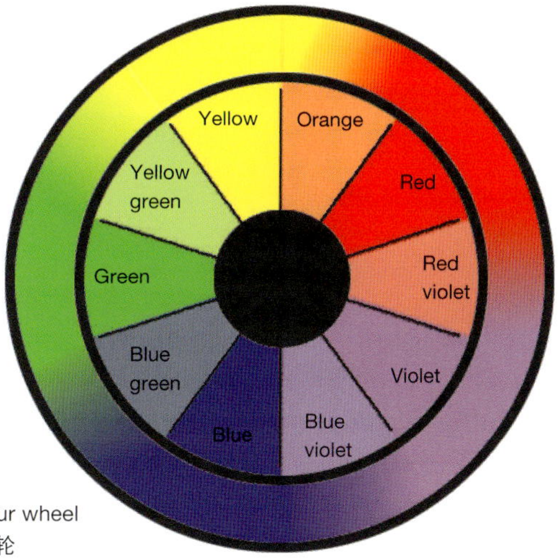

Colour wheel
颜色轮

5.4

- It is recommended natural lighting shall be provided in all bedrooms, living / dining rooms, kitchens and toilets.
- It is recommended to avoid source of glare, such as large windows at the end of dark corridors.
- It is recommended that natural lighting shall be provided in entrance lobbies to provide a transition between outdoor and indoor illumination levels.
- It is recommended that directional task lighting shall be provided at workplaces, such as food preparation areas in kitchens or reading places in bedrooms or living rooms. It is desirable that the position and illumination level of task lighting can be adjusted by users themselves.
- It is recommended to provide spot lighting to room signages, keyholes and handles at flat entrances.
- To facilitate use by people with visual impairments, it is recommended to minimise overall visual congestion.

- 建议所有卧室、客/饭厅、厨房和厕所都使用自然采光。
- 建议须避免眩光，如避免在漆黑的走廊末端设置大型窗户。
- 建议在入口大堂提供自然采光，作为室外照明度与室内照明度的过渡区。
- 建议在工作地点应设置指向工作灯，例如在厨房备餐台或在卧室或客厅的阅读处，并最好应能自行调校工作灯的位置和照明。
- 建议在楼层入口设置照明灯，以照亮门牌、匙孔和扶手。
- 建议应尽量减少视觉上的负荷，以方便视障人士。

A large window at the end of a dark corridor is a source of glare
在漆黑走廊末端的大型窗户会引至眩光

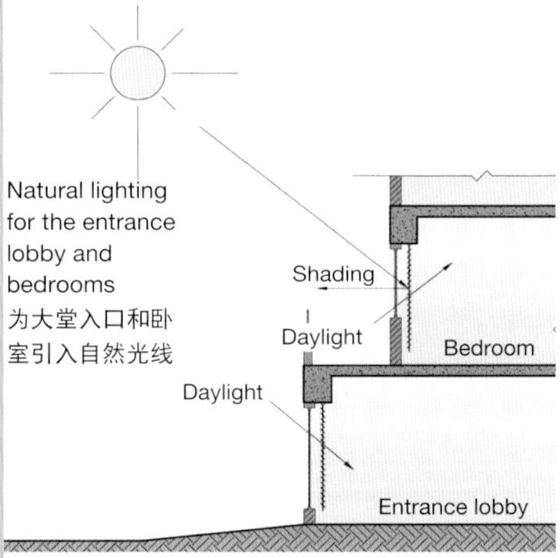

Natural lighting for the entrance lobby and bedrooms
为大堂入口和卧室引入自然光线

5.5 ACOUSTIC ENVIRONMENT

The elderly and the weak have a high sensitivity to noise in the living environment. To mitigate environmental noise, it is recommended that:

- Residential development shall be located away from busy vehicular traffic.
- Trees shall be planted around buildings to act as noise barriers.
- Acoustically absorbent materials shall be used for walls and floors.
- Internal walls shall have acoustic proof level of greater than 40dB.
- Hard floor surfaces that produce echoes shall be avoided.
- Alarm signal shall be no less than 65dB. However, this can be reduced to 60dB in enclosed staircases and enclosures of no more than 60m^2 in area.
- Where background noise is more than 60dB, the alarm signal shall be increased to be 5dB more than background noise.
- Alarm signals in bedrooms should be not less than 75dB.
- Where ambient noise in the external environment exceeds 60dB, double-glazed windows shall be provided to achieve 40dB or less.

5.5 听觉环境

长者和体弱人士很需要有宁静的居住环境。为减低噪声干扰，建议实行下列措施：

- 居住建筑应远离繁忙的交通。
- 在楼宇周边种植树木，作为隔声屏障。
- 墙和地板均使用隔声物料。
- 内墙的隔声量应多于40dB。
- 避免使用会发出回音的硬地板。
- 警报器音量一般要达到65dB以上；然而在密封的楼梯或围封范围少于60m^{2}的地方，则可减至60dB。
- 周遭噪声如高于60dB，警报响闹应较周遭噪声量高出5dB。
- 卧室的警报器声量不低于75dB。
- 如室外噪声声响高于60dB，则应安装双层玻璃窗户，使噪声减至低于40dB。

5

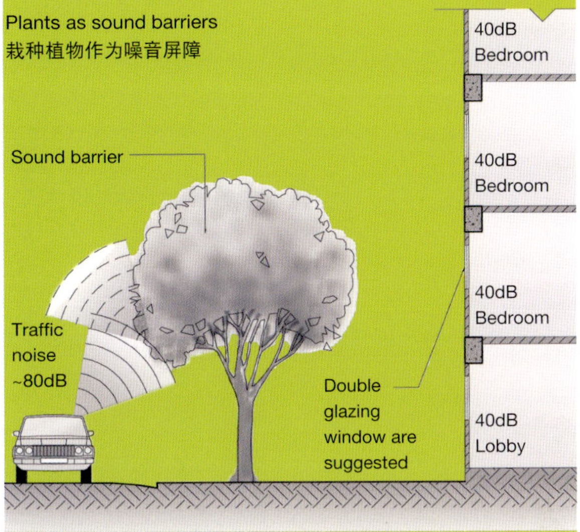

5.6 生活空间

生活空间的大小会影响居民的心理和健康，特别是那些长时间留在家中的人。净空高度和室内面积是两大重要因素。今建议如下:

- 客/饭厅和卧室的净空高度应不少于2500mm。
- 房间面积不少于下表所示:

房间	建议最小面积 [m²]
客/饭厅	14 （3.5m x 4m）
单人房	5 （2m x 2.5m）
双人房	9.5 （2.7m x 3.5m）
主人房	11 （2.9m x 3.8m）
厨房	4 （1.8m x 2.7m）
有淋浴间的厕所	3.3 （1.8m x 1.8m）

5.6 LIVING SPACES

The dimension of living space can affect the psychological health and well being of residents, particularly those who spend long periods at home. Headroom and room areas are two essential factors. It is recommended that:

- The clear headroom of living / dining rooms and bedrooms shall be not less than 2500mm.
- Minimum room areas shall be as follows:

Room	Recommended minimum areas [m²]
Living / dining room	14 （3.5m x 4m）
Single-bed room	5 （2m x 2.5m）
Double-bed room	9.5 （2.7m x 3.5m）
Master bedroom	11 （2.9m x 3.8m）
Kitchen	4 （1.8m x 2.7m）
Toilet with shower	3.3 （1.8m x 1.8m）

5.7 水管及渠道

- 下水道口应有水封满。
- 建议在洗涤盆、洗面盆、淋浴间和浴缸附近装置容易清理垃圾的下水道盖。
- 建议应清楚标示出水弯管和水龙头。
- 建议应定期检查所有出水弯管和渠道，确定是否渗漏，并妥善维修。

5.7 PLUMBING AND DRAINAGE

- Drainage traps should be always kept sealed with water.
- It is recommended that easily cleaned and operated waste traps shall be provided to sinks, basins, showers and bathtubs.
- It is recommended that valve and taps / faucets shall be clearly labelled.
- It is recommended that regular inspection shall be undertaken to identify leaking pipes and proper maintenance of all plumbing and drainage fittings and fixtures.

Easily cleaned floor drain fittings for showers/ bathrooms
容易清理的浴室地漏配件

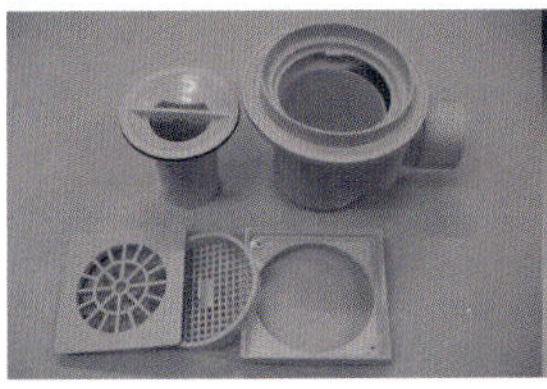

Easily cleaned drain fitting for sinks
容易清理的洗涤盆下水配件

章
Cha

6

pter

Good Practices on Detailing
装饰详细设计的推荐方案

6.1

6.1 GOOD PRACTICE FOR FLOOR FINISHES

6.1.1 Comparison of Floor Finishes

Floor finishes affect accessibility, the acoustic environment, the thermal environment, health and safety as well as capital cost and maintenance. Some advantages and disadvantages of popular flooring materials are considered below:

Flooring	Advantages	Disadvantages
Carpet	• Soft • Slip-resistant • Reduce the impact of falls	• Easy to get stained • Trap large quantities of dust • Not resistant to moisture
Vinyl / Linoleum	• Soft • Easily installed • Reduce the impact of falls • Low cost • Wide range of patterns and colours	• Emit gases with high levels of chemicals, particularly plasticizers when it is new
Linoleum	• Soft • Easily installed • Reduce the impact of falls • Low cost • Wide range of patterns and colours	• Emit a strong odour that may cause respiratory irritation
Ceramic tile	• Suitable for wet rooms • Stain resistant • Durable • Scratch resistant • Not emit toxic fumes • Not retain odours, allergens or bacteria	• Hard • Slippery when wet, slip-resistant coating should be introduced
Timber flooring	• Durable • Warm and home-like	• Not resistant to moisture • Not soft • Easily scratched • Expensive
Natural stone	• Durable	• High reflectiveness • Slippery when wet • Cold surface • Brittle

6.1 地板设计的推荐方案

6.1.1 地板物料比较

地板物料会影响畅达程度、听觉环境、温度、安全、成本和维修。现时普遍采用的地板物料有下列各种，其优劣分析如下:

地板物料	优点	缺点
地毯	• 软身 • 防滑 • 降低跌伤的后果	• 容易玷污 • 存留大量尘埃 • 不能抵御潮湿
胶地板/漆地胶地板	• 软身 • 容易铺设 • 降低跌伤的后果 • 价格便宜 • 款式颜色多样化	• 新制成的地板会释出含大量化学品的气体、尤其是增塑剂
漆地胶地板	• 软身 • 容易铺设 • 降低跌伤的后果 • 价格便宜 • 款式颜色多样化	• 释出强烈气味、可能引至呼吸道过敏
瓷砖	• 适用于潮湿的房间 • 不易沾染污渍 • 耐用 • 不易出现刮痕 • 没有毒气释出 • 不留气味、敏感源或细菌	• 坚硬 • 弄湿后易使人滑倒，宜加上防滑层
木地板	• 耐用 • 有温暖和家居感觉	• 不耐潮湿 • 不软身 • 容易有刮痕 • 昂贵
天然石块	• 耐用	• 反光 • 弄湿后易使人滑倒 • 表面冰冷 • 易碎

6

装饰详细设计的推荐方案

6.1

Wall to wall vinyl and linoleum sheets are recommended for a wide range of applications. Both are easily cleaned and have a resilient surface, which slightly mitigates fall injury. Highly polished natural stones are not recommended, because reflections may cause confusion to the elderly. Furthermore, their hard and slippery surface can easily cause fall injury.

建议采用胶地板和漆地胶地板。这两种物料方便清洁，加上富弹性的表面，可轻微减轻跌伤的程度。高度磨光的天然石容易反光，容易使长者产生混淆；加上坚硬及易使人滑倒的表面容易引至跌伤，故不建议使用。

6.1.2 Details of Floor Finishes
地板物料详细设计

Vinyl / Linoleum Floor Finishes
胶地板/ 漆地胶地板的物料

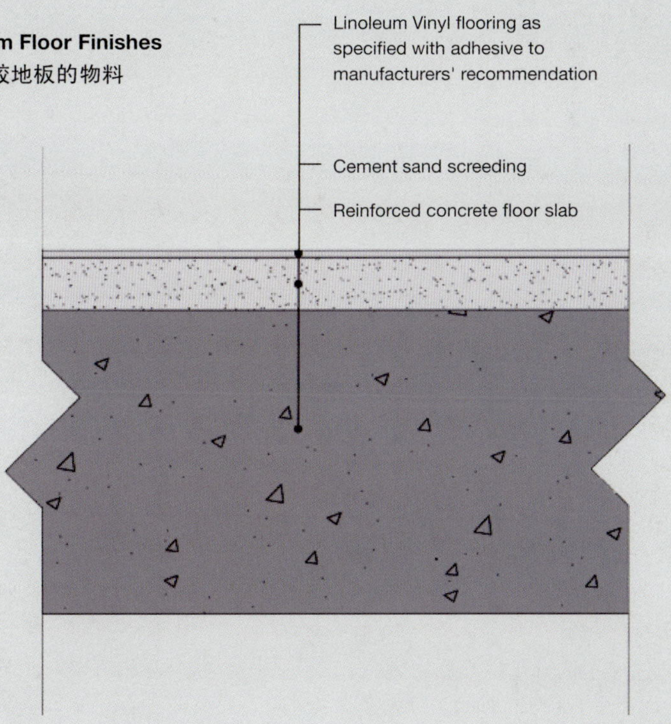

Vinyl skirting

To prevent dust collection, three types of vinyl skirtings are shown:

胶地板的墙裙

避免存留尘埃，可以使用下列三种胶墙裙:

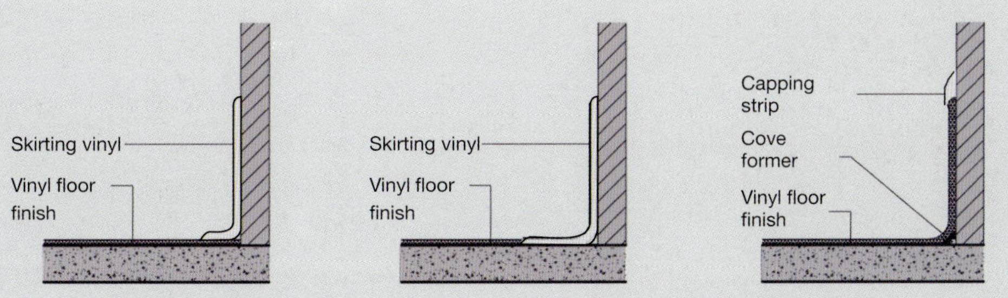

Tactile Tiles
引路砖

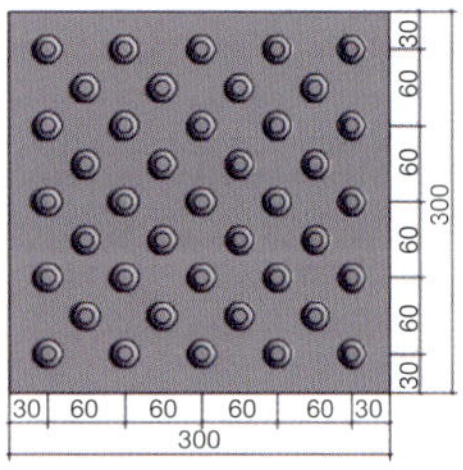

Positional tile/block
方位引路砖

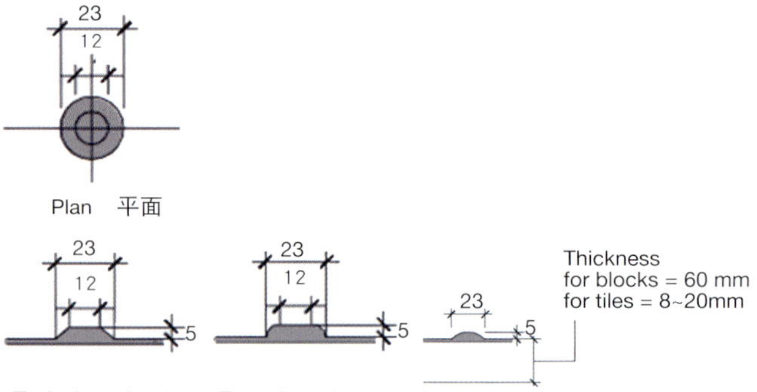

Typical section 1
标准剖面 1

Typical section 2
标准剖面 2

Typical section 3
标准剖面 3

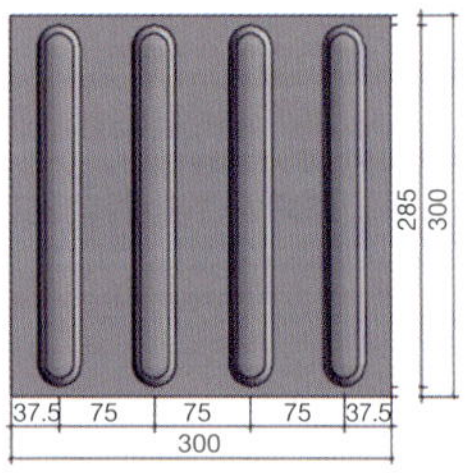

Directional tile/block
方向引路砖

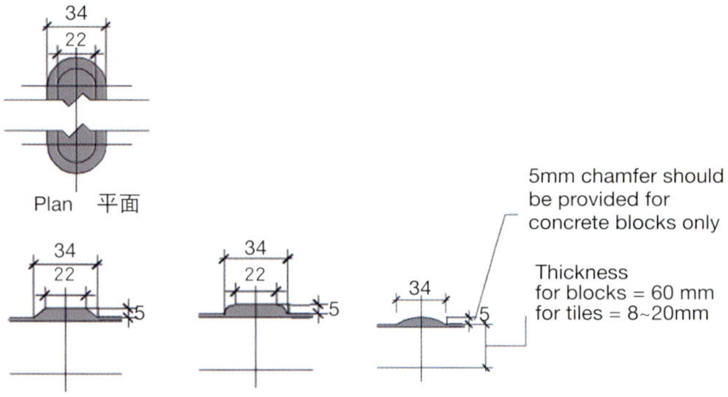

Typical section 1
标准剖面 1

Typical section 2
标准剖面 2

Typical section 3
标准剖面 3

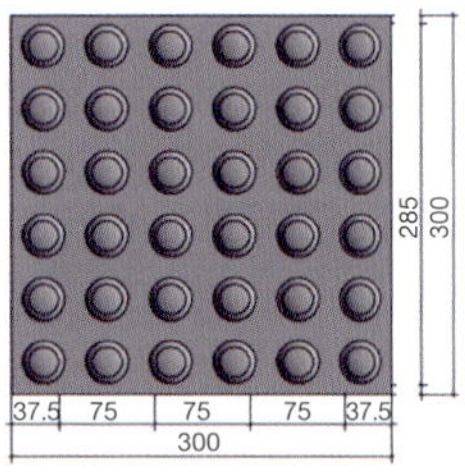

Hazard warning tile/block
危险警告引路砖

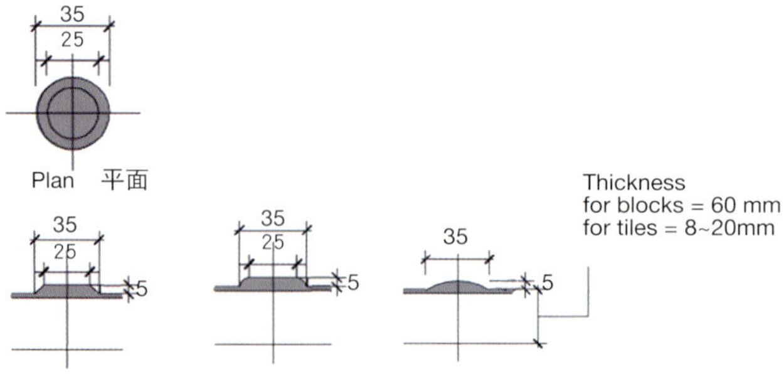

Typical section 1
标准剖面 1

Typical section 2
标准剖面 2

Typical section 3
标准剖面 3

装饰详细设计的推荐方案

6.2 GOOD PRACTICE FOR THRESHOLDS

6.2

Thresholds have traditionally been provided to prevent water and blown dirt from entering a property. To facilitate access by the elderly and people in wheelchairs, it is recommended that door thresholds shall not exceed 15mm in height and shall be levelled. Ideally, flush thresholds are suggested to prevent tripping, fall injuries and improved accessibility. To achieve accessible thresholds, good practice is introduced as follows:

6.2.1 Levelled Thresholds at Building Entrances

Where a levelled threshold to a building entrance is proposed, drainage channelwith a suitable channel cover should be provided to prevent the ingress of water.

6.2 门槛设计的推荐方案

传统上门槛用来防止水和尘垢进入屋内。为方便长者及轮椅人士出入，建议门槛的高度不应高于15mm，并要切成斜面。最理想是使用平的门槛，可防止绊倒跌伤，并可改善畅达程度。无障碍门槛的推荐方案列举如下:

6.2.1 楼宇入口门槛

住宅入口处若设齐平门槛， 应设排水设施以阻挡水进入室内。

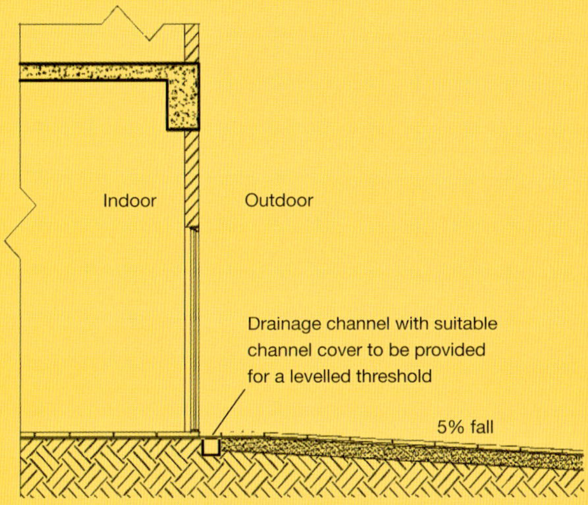

Entrance of levelled thresold at a building entrance
楼宇入口的齐平门槛样本

Levelled threshold at a building entrance
楼宇入口的齐平门槛

6.2.2 **Door Thresholds at Kitchens and Toilets/Bathroom**
厨房和厕所/浴室门槛

Good practice for renovation of thresholds at toilets and kitchens
厕所/浴室和厨房门槛的推荐改建方案

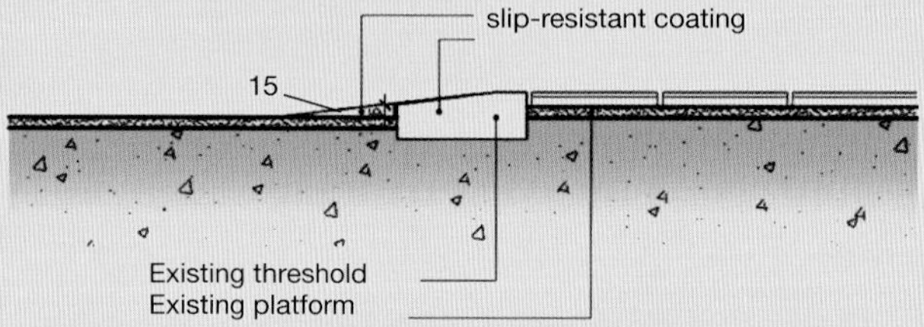

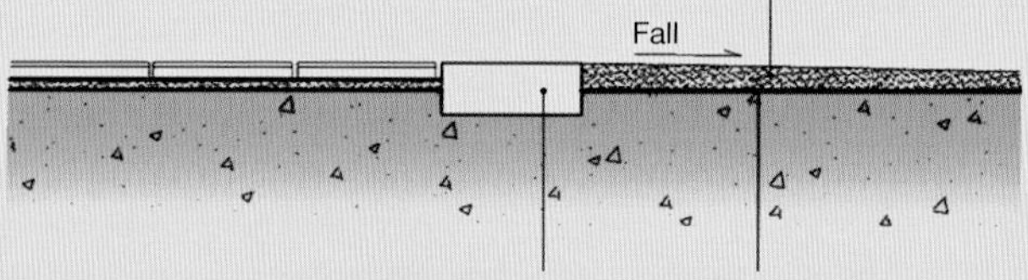

The threshold at the toilet is too high
淋浴室齐平门槛的样本

Good practice for thresholds at toilets and kitchens
淋浴室齐平门槛的样本

Example of levelled thresholds for kitchens and toilets
厨房/厕所齐平门槛的样本

6

装饰详细设计的推荐方案

6.2

6.2.3 Thresholds at Balconies
露台门槛

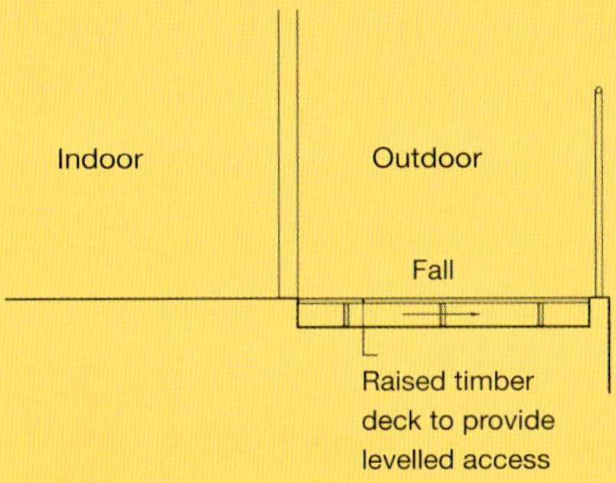

Good practice for thresholds at balconies
露台门槛的推荐方案

Bad practice for thresholds at balconies
露台门槛太高的不良做法

6.2.4 Entrance Mats

To facilitate access by people in wheelchairs, it is recommended that entrance mats shall be secured to the floor surface. Rubberized "horsehair" or coir coconut types of entrance mats are not recommended, as they make wheeled traffic difficult. However, mats that consists of aluminium channels with coir fibres or carpet pieces mounted in the channels can provide a firm surface to wheeled traffic and thereby enable accessibility for people in wheelchairs.

6.2.4 楼宇入口地垫

为方便轮椅人士出入，建议入口地垫最好能紧贴地面，不应使用塑胶「马毛」和椰衣类别的入口地垫，因会妨碍轮椅的转动；但可使用有结合椰衣纤维或地毯块及铝条的地垫，其表面坚固，有助滑轮的活动，方便轮椅人士使用。

Bad practice - the entrance mat is not secured to the floor surface
门口地垫不紧贴地面的不良做法

6

6.3 GOOD PRACTICE FOR DOOR HANDLES

To facilitate use by the weak and people with impaired dexterity, it is recommended that lever door handles shall be preferred to round knobs, as they are easier to operate. People with arthritic hands can use their elbows or the edge of their hands to operate lever handles.

To prevent users catching clothing such as pockets, it is recommended that lever handles shall have their outer end return towards the door.

6.4 GOOD PRACTICES FOR DOORS

6.4.1 Vision Panels

Vision panels shall be provided to kitchen doors, and to doors in public areas, such as to staircases. It is recommended that the bottom edge of a vision panel shall be at a height of not more than 1000mm and the top edge at a height of not less than 1500mm above floor level.

6.3 门把手设计的推荐方案

由于杠杆式门把手较球形把手更容易操控，故建议使用杠杆式门把手，以方便体弱及手脚不灵巧的人士使用。手部关节有毛病的人士，可使用手肘或手部的边缘来操控杠杆式门把手。
为防止使用者的衣服例如衣袋给钩住，杠杆式门把手的外端应回转向门的方向。

6.4 门设计的推荐方案

6.4.1 观望板

厨房门和通往公众地方的门，例如楼梯门应设置观望板。建议观望板底部离地面的高度不应高于1000mm，顶端离地面高度则不应少于1500mm。观望板设计的推荐方案，见右附图。

6.3

Good practice for lever handles
手柄形把手的推荐方案

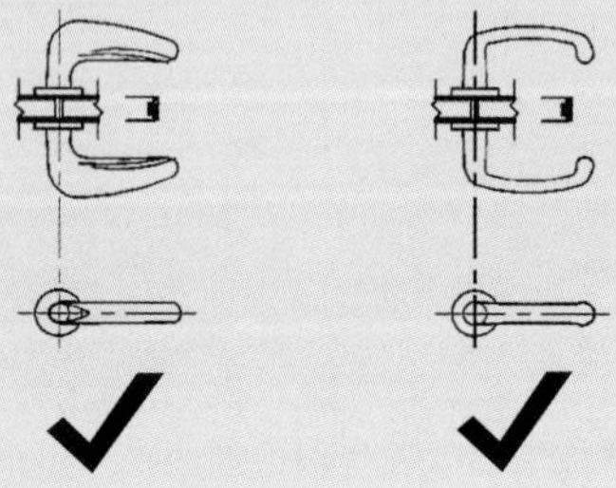

6.4

Good practice for vision panels
观望板的推荐方案

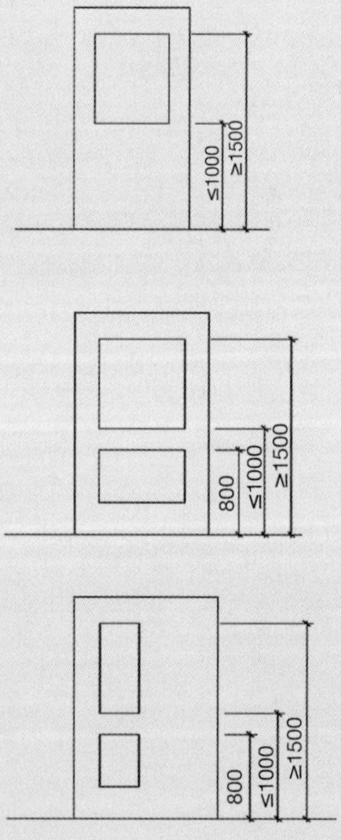

Good practice for keyholes
匙孔的推荐方案

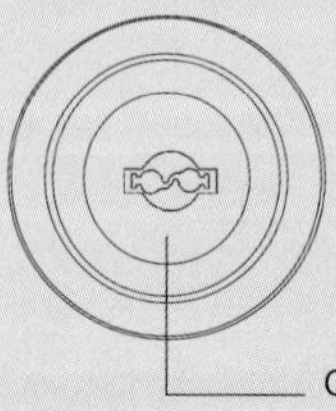

装饰详细设计的推荐方案

6.4

6.4.2 Spy-holes

To facilitate use by children and people in wheelchairs, it is suggested that two spy-holes can be provided at different heights. The lower one shall be at a height of between 750mm and 850mm. The higher one should be at a height of between 1350mm and 1500mm.

6.4.3 Keyholes

To facilitate use by people with visual impairments, it is recommended that keyholes shall be provided which are distinguished by contrasting colours. Keyways with concave cylinder face and reversible keying system can help people locate the position of keyholes.

6.4.4 Sliding Doors

To facilitate use by persons of limited mobility and people in wheelchairs, it is suggested that sliding doors can be provided for small rooms, such as pantries, toilets and bathrooms.

6.4.2 防盗眼

为方便儿童及轮椅人士使用，建议于不同高度可设置两个防盗眼，较低者离地面高度应介乎750～850mm，较高者离地面高度应介乎1350～1500mm。

6.4.3 匙孔

为方便视障人士使用，建议用对比色来分辨匙孔的位置。有凹入圆锥形面的钥匙槽和可倒置插匙的锁，有助找出匙孔的位置。

6.4.4 推拉门

为方便活动能力有限及轮椅人士使用，建议备餐室、厕所和浴室等小面积的房间应使用推拉门。

Different types of sliding doors are shown as follows:
各种不同的推拉门列举如下:

Single-leaf sliding doors with straight track
行直轨的单扇推拉门

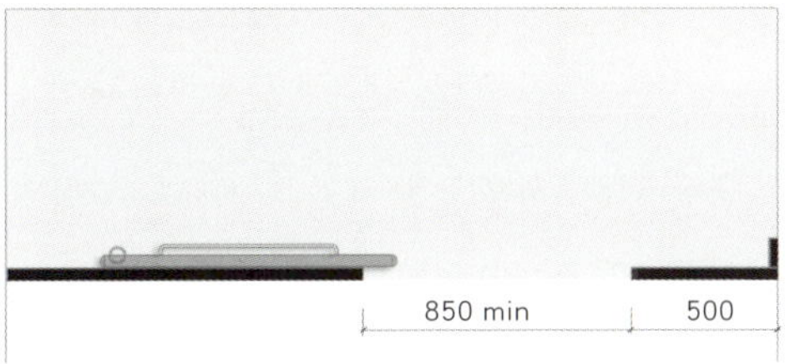

Plan of single-leaf sliding doors with curve track
行曲轨的单扇推拉门平面图

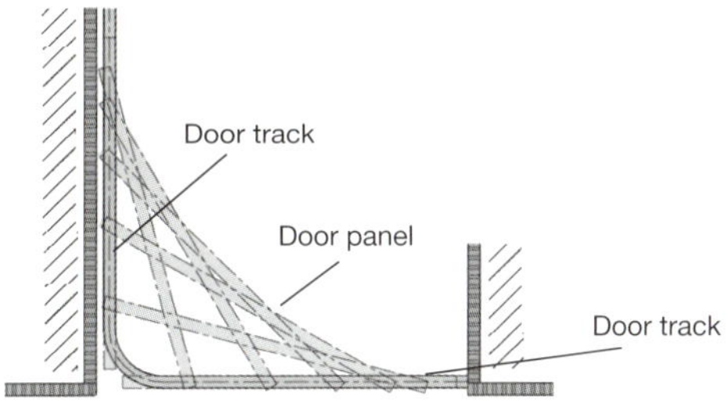

Double sliding doors
双扇推拉门

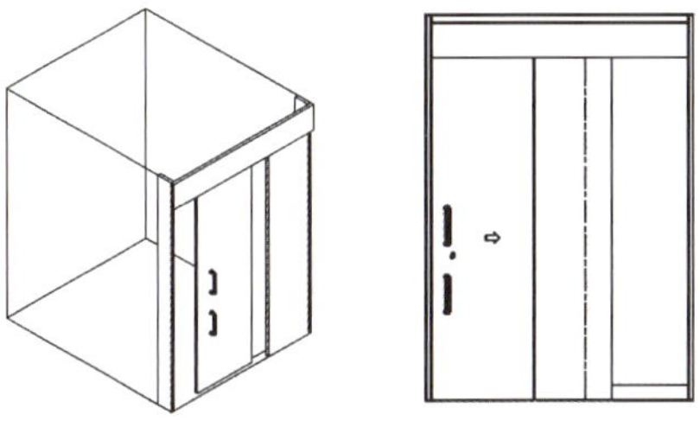

Design of two spy-holes for flat entrance doors
楼宇门口有两个防盗眼的设计

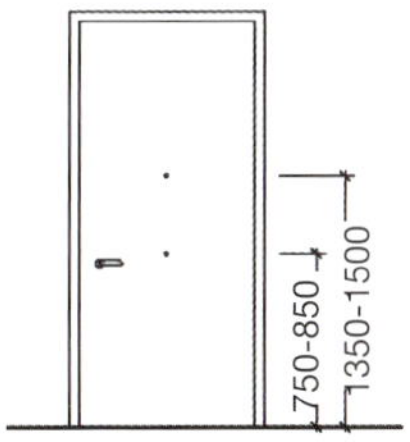

Two spy-holes to facilitate use by children and people in wheelchairs
有两个防盗眼以方便儿童和坐轮椅人士使用

Example of design of two spy-holes for flat entrance doors
楼宇门口有两个防盗眼的设计样本

6.5

6.5 GOOD PRACTICE FOR HANDRAILS

6.5.1 Materials for Handrails

It is suggested that handrails can be provided with a nylon coated finish. Nylon has a wide colour range and is comfortable in use. To provide extra grip, it is suggested that grooved patterns shall be provided on the wall-side surface.

6.5.2 Details of Handrails

Handrails Mounted in Wall Recess
设于墙壁凹位的扶手

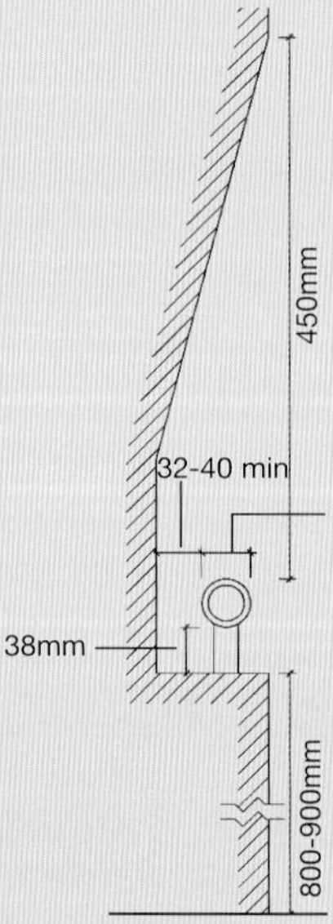

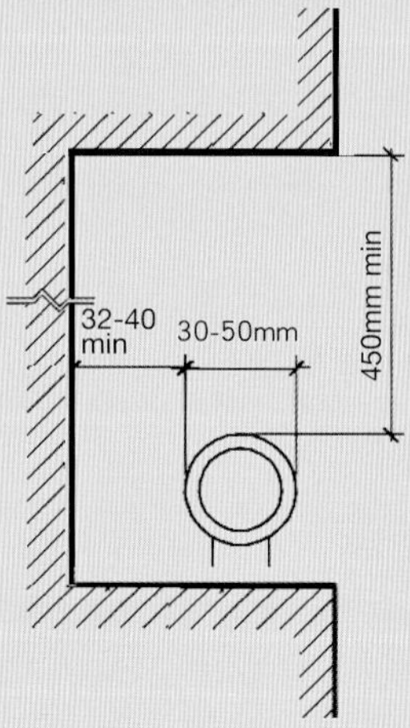

6.5 扶手设计的推荐方案

6.5.1 扶手物料

建议扶手可铺设尼龙面层。尼龙物料的颜色众多，而且舒适。为使扶手在使用时更加稳固，建议应于近墙一边加上槽纹。

6.5.2 扶手详细设计

Handrails Mounted on Wall Surface
设于墙面的扶手

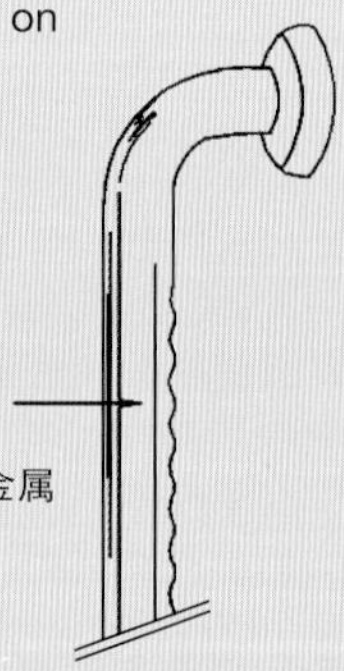

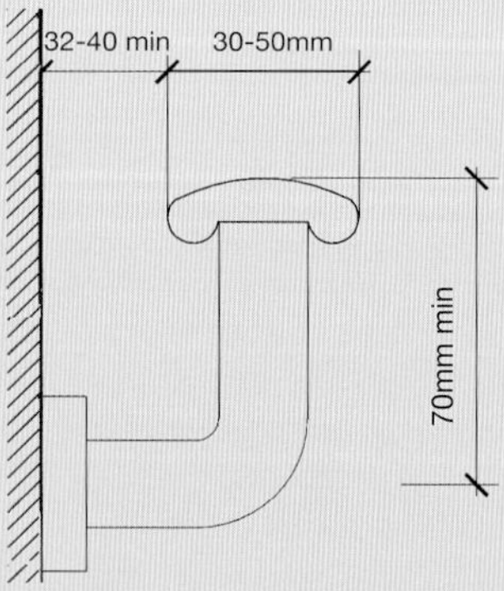

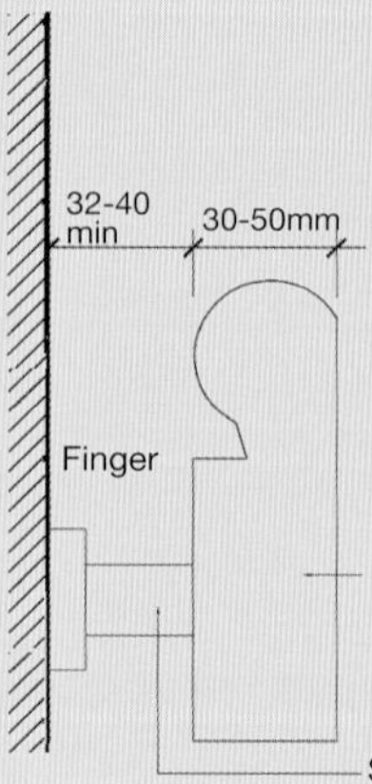

6

装饰详细设计的推荐方案

6.5

6.5.3 门边扶手

为方便长者和体弱的人士使用，建议在门的旁边加上垂直扶手，尤其是一些要大力开启的门户，其可提供一个支力。建议垂直扶手应离门边150～200mm，扶手的底部离地面不可高于850mm。

6.5.3 Handrails at Doors

To facilitate use by the elderly and the weak, it is recommended that a vertical handrail shall be provided beside doors. This provides a brace when the opening force is high. It is recommended that the vertical handrail shall be provided 150mm ~ 200mm away from the door, with its bottom at a height of not more than 850mm above floor level.

6.5.4 住宅入口扶手

为方便视障人士使用，建议应在门口的旁边设向上弯曲的扶手。这既可提示门口的位置，又为长者和体弱的人士提供一个支架，以方便他们开门。

6.5.4 Handrails at Flat Entrances

To facilitate use by people with visual impairments, it is recommended that handrails with upturns shall be provided beside entrance doors. This can indicate the locations of openings and doors. It also provides a brace to assist the elderly and the weak to open the doors.

6.5.5 橱柜扶手

为方便长者和体弱的人士使用，建议有铰接门的橱柜应安装横向的扶手，可帮助使用者更方便开门。

6.5.5 Handrails on Cabinets

To facilitate use by the elderly and the weak, it is recommended that horizontal handrails shall be provided where cabinets have hinged doors. This can assist users to open the doors.

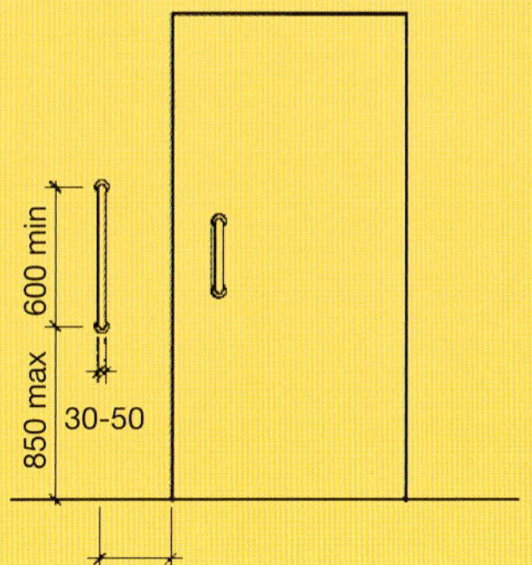

Handrails at doors

门边扶手

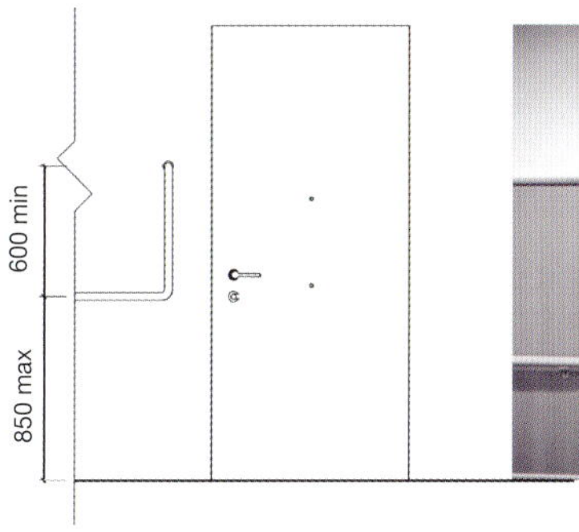

Handrails with upturns at entrance doors

向上屈曲的扶手

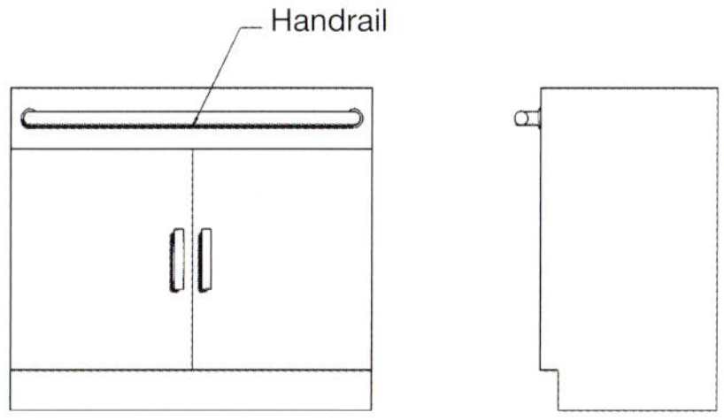

Handrails on cabinets

厨柜扶手

6.6 GOOD PRACTICE FOR SIGNAGE

Residential design encompasses a variety of building types: At one extreme these range from traditional single storey houses where users know where everything is, how it works and may have made it themselves. At the other extreme are extensively managed high-rise facilities where users come from a variety of cultural backgrounds and may need guidance on how the building is intended to be used. For traditional housing, signage is hardly an issue. For innovative housing it can be essential to sustainable operation.

6.6 指示牌设计的推荐方案

住宅设计包括各色各样的建筑物类型：传统的单层房屋，居住者知道每一物件的所在，如何运作，房屋还可能是他们自己盖的；另一方面，需要广泛管理的高层建筑物设施，居住者来自不同的文化背景，可能需要指示该建筑物是预定如何使用的。对传统的房屋，缺少指示牌并不是一个问题。对设计创新的房屋，如要持续运作，指示牌必不可少。

6.6.1 International Symbols

Examples of international symbols for users with disabilities:

6.6.1 国际标志

与残障人士相关的国际标志举列如下：

Symbol to indicate accessible parking

Symbol to indicate an independently accessible entrance

Symbol to indicate a ramped entrance

Symbol to indicate level access throughout the building

Symbol to indicate that assistance is available

Symbol to indicate wheelchair accessible toilet

Symbol to indicate wheelchair accessible lift

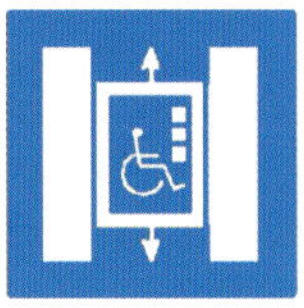

Symbol to indicate wheelchair accessible telephone

Symbol to indicate an introduction loop

Symbol to indicate an infra-red system

6.6.2 Directional Signage for Wayfinding

In complex buildings there may be circumstances where the layout is not easily self-explanatory. Where this occurs it may be necessary to introduce directional signage, particularly at locations where the user has to make a decision.

Two types of strategies can be used for wayfinding design, namely architectural and information wayfinding design. These two types of strategies can reinforce and complement each other.
Examples are shown as follows:

6.6.2 方向指示牌

在结构复杂的楼宇，在某些情况下，平面布置不容易被识别。如有这样的情况发生，便可能需要使用方向指示牌，特别是当使用者须要确定方向时。

方向指示标志的设计主要运用两种设计手法，即建筑设计型及信息供应型两种方向指示设计手法。此两种设计手法相辅相承。部分例子举例如下:

Signage at junction wall
墙角指示牌

Information board near the entrance
入口附近的资料查询设施

Information board
方位指示牌

Directional signages for entrances
入口方向指示牌案例

Electronic information board
电子资料查询设施

Floor patten and wall signage to help identify flat entrances
铺地图案及门牌设计可帮助释别住宅入口方位

6.7 GOOD PRACTICE FOR KITCHENS
厨房设计的推荐方案

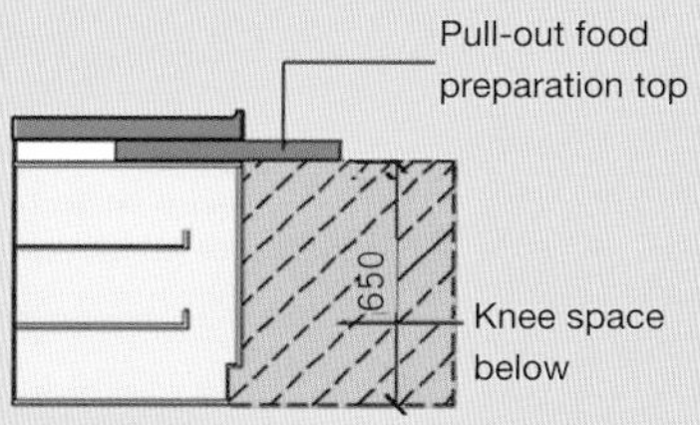

Pull-out food preparation tops
可拉出的煮食台

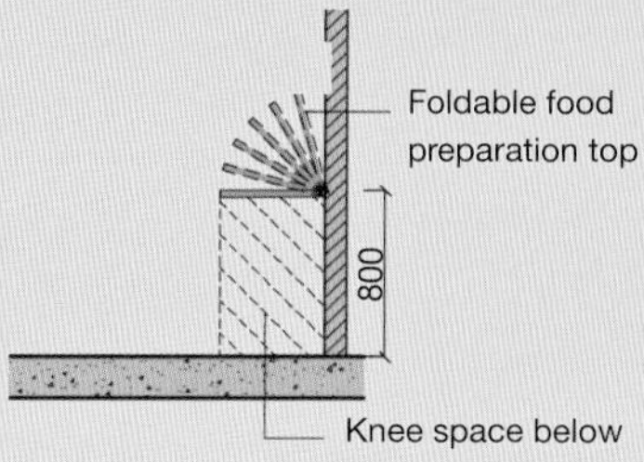

Foldable food preparation tops
可折叠的煮食台

6.7

6.7.1 Food Preparation Tops for Kitchens

To facilitate use by people in wheelchairs, it is recommended that knee space shall be provided under food preparation tops. Where this is difficult to achieve, a foldable or pull-out food preparation tops can be provided for light-weight use.

6.7.1 厨房的煮食台面

建议在煮食台面下预留膝盖空间。如实践上困难，可使用可折叠或拉出柜台，以方便轮椅人士进行简单和不大用力的工序。

6.7.2 Knee Spaces for Kitchens

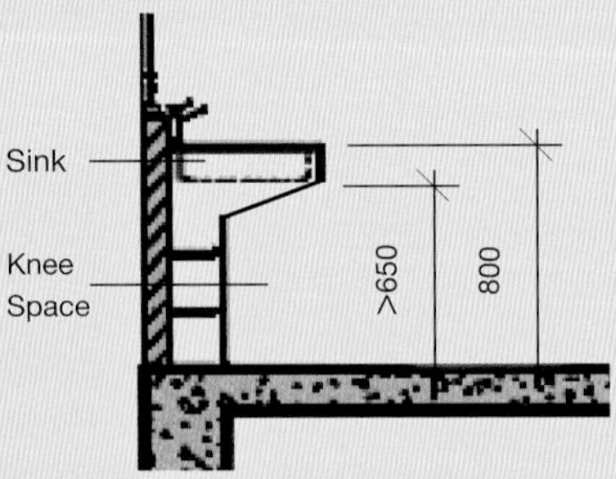

Knee space under the sink
洗涤盆下面的膝部空间

6.7.2 厨房的膝盖空间

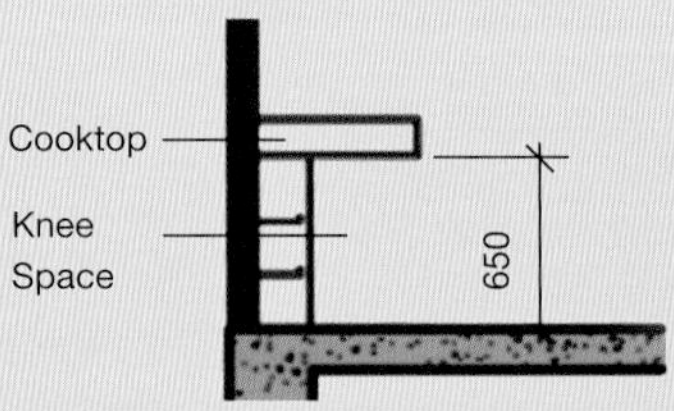

A knee space under the cooktop
煮食柜下面的膝部空间

6.7.3 Electrical Sockets for Kitchens

When an electrical socket outlet is located above a kitchen counter, the bottom edge should be at a height of not less than 75mm from the top of the counter, ideally not more than 1100mm above floor level.

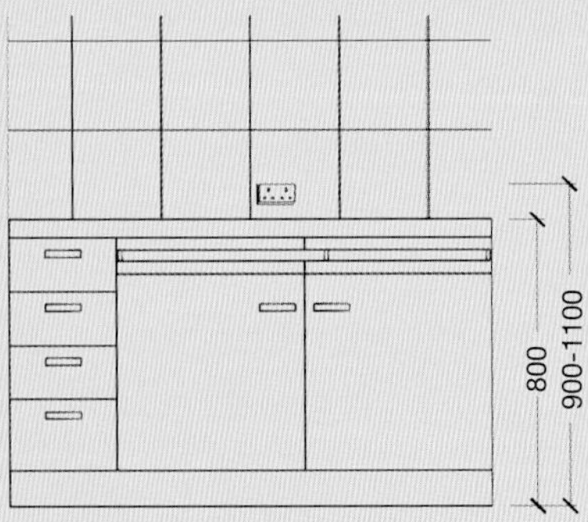

Location of electrical socket outlets for kitchen
厨房插座位置

6.7.3 厨房的插座

插座若是安装在厨房柜台之上，其底边离柜台表面至少应有75mm，其高度离地面以不超过1100mm为理想。

6

装饰详细设计的推荐方案

6.7

6.7.4 厨房家具

厨房柜台面的高度不应超过800mm，以方便轮椅人士。悬于墙壁橱柜的底部也不应高于1200mm。

为要提供更多贮物空间，同时在厨房柜台下提供膝盖空间，可以装置可移动的橱柜。只要结构稳固、高度合宜、滚轮能够稳妥刹停，可移动的橱柜可作座位使用。

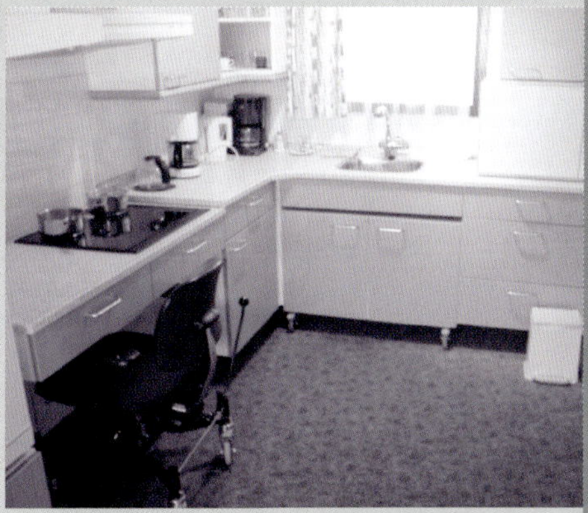

Good practice for kitchen furniture design
厨房家具的推荐方案

6.7.4 Kitchen Furniture

To facilitate use by people in wheelchairs, it is suggested that kitchen counters shall have a height of not more than 800mm. The bottom of wall-hung cabinets shall be at a height of not more than 1200mm.

In order to provide more storage space and also to achieve knee space under the kitchen counter, movable cabinets can be provided. Movable cabinets can also be used as a seat, provided that construction is sufficiently robust, the height is suitable and the wheels include brakes.

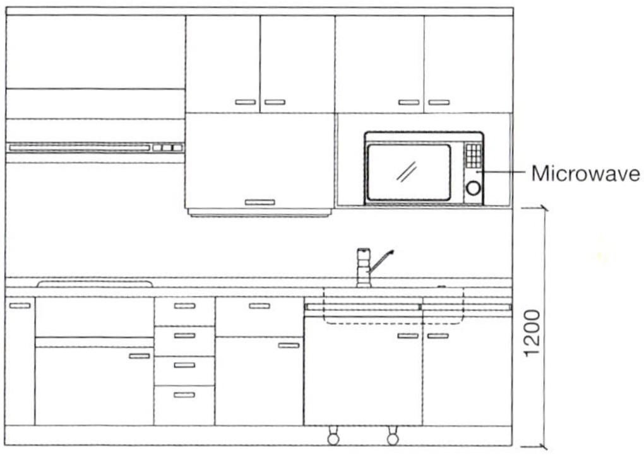

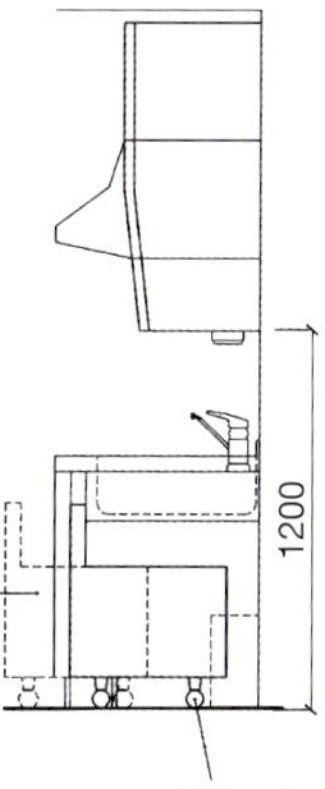

Example of kitchen furniture designs
厨房家具样本

6.8 GOOD PRACTICE FOR TOILETS, BATHROOMS AND SHOWERS

6.8 厕所、浴室及淋浴间设计的推荐方案

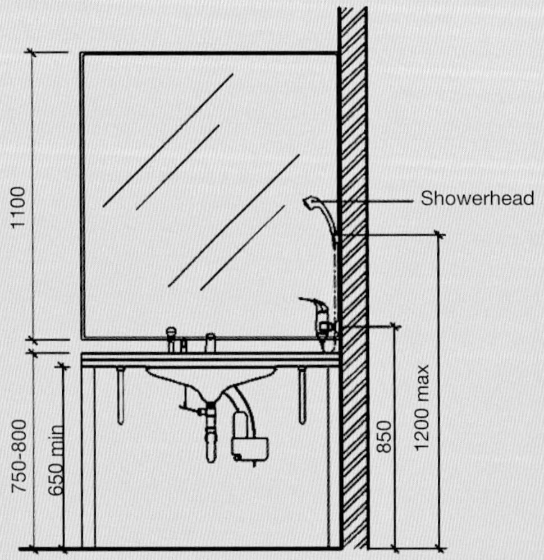

Showerhead is installed at a basin
装在洗涤盆上的花洒头

6.8.1 Showerheads

- It is suggested that a showerhead should be installed at a basin for hair washing.
- It is recommended to provide flexible hoses and sliding rails so that the height of showerheads can be adjusted.
- At basin, it is possible to provide a tap/faucet with a pull-out flexible hose for hair washing.

6.8.1 花洒头

- 建议在洗面盆上安装花洒头，以方便洗头。
- 建议安装软身的喉管和滑轨，以方便调校花洒头高度。
- 可在洗面盆上安装喉管可以拉长的水龙头，以方便洗头。

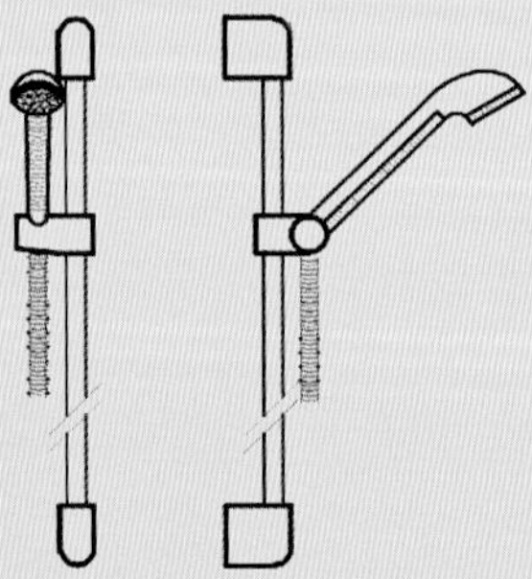

Height adjustable showerhead
可调校高度的花洒头

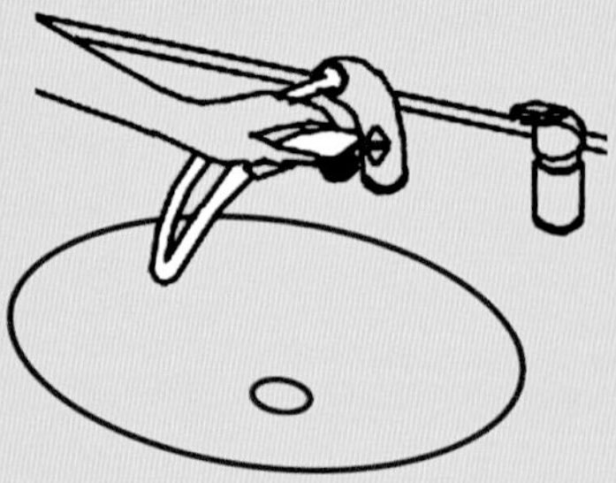

A faucet with a pull-out flexible hose
可拔水龙头

6.8.2 Shower Seats

It is recommended that the height of shower seats shall be adjustable. Different types of shower seats are shown:

6.8.2 淋浴椅

建议淋浴椅的高度应可调校，下面列举三种淋浴椅的款式:

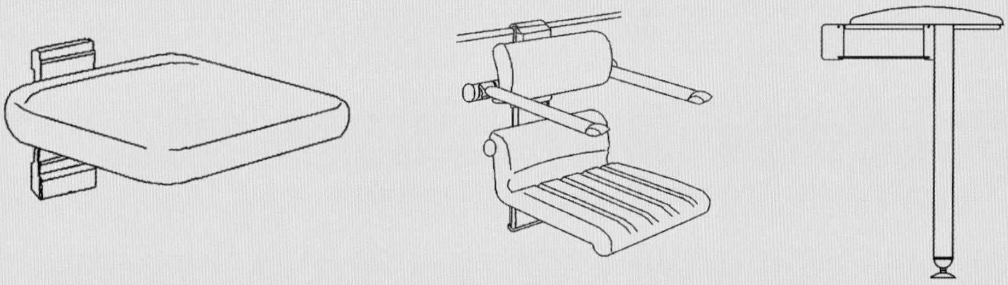

6.8.3 Wash Basins

To facilitate use by people in wheelchairs, it is recommended that knee space shall be provided under basins. Where storage is required, shallow cabinets can be provided in combination with semi-countertop basins.

6.8.3 洗面盆

建议洗面盆下面应预留膝盖空间。如需要有储存空间，则可使用浅身的橱柜，配合半柜栅面的洗面盆，以方便轮椅人士。

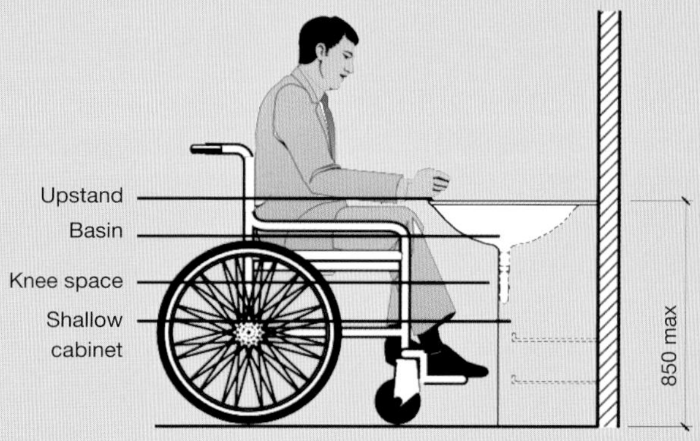

Wash basin with knee space
设有膝盖空间的洗面盆

6.8.4 Bathtubs

To facilitate use by the elderly and people with disabilities, it is recommended that bathtubs shall be provided with transfer platforms. Three types of transfer platforms / boards are shown:

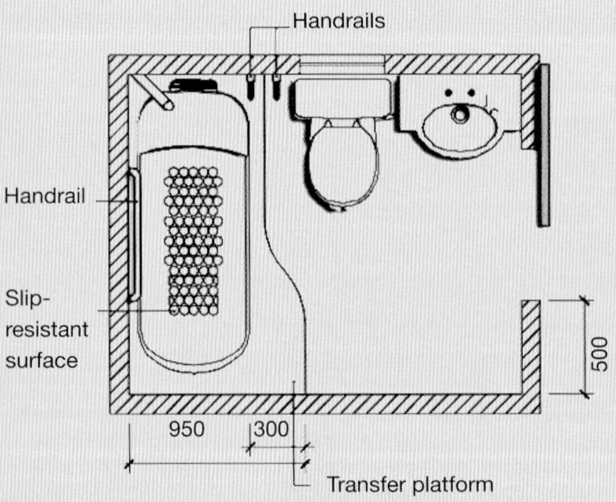

Option 1
方案 1

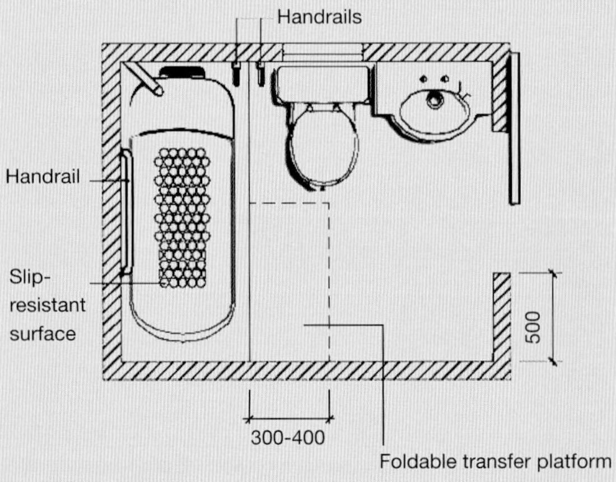

Option 2
方案 2

6.8.4 浴缸

建议使用有转移平台板的浴缸，方便长者或残障人士使用。下面列举有三种转移平台板的款式:

A bathtub with a bathtub board
设有浴缸板的浴缸

Transfer platforms can also be utilised as a location for a freestanding baby bath.
转移平台可以用作摆放婴儿的沐浴盆。

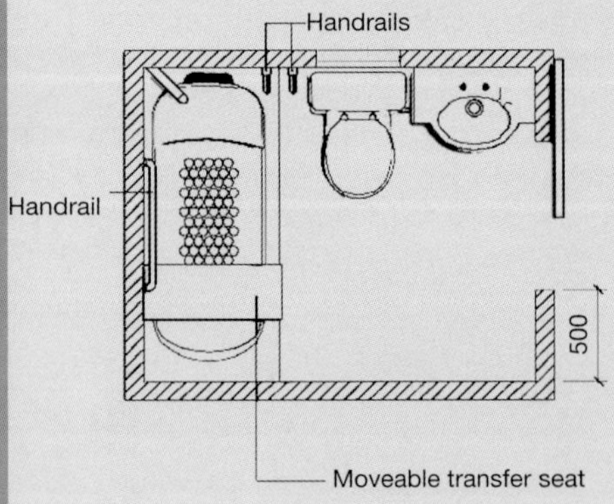

Option 3
方案 3

6.8.5 Shower Floor Grilles

To facilitate assisted showering in an accessible shower, it is suggested that shower floor grilles can provide a level and drained surface.

6.8.5 淋浴地面栏栅

建议在淋浴地面使用栏栅，提供齐平地面和去水平面，以方便淋浴时需要帮助的使用者。

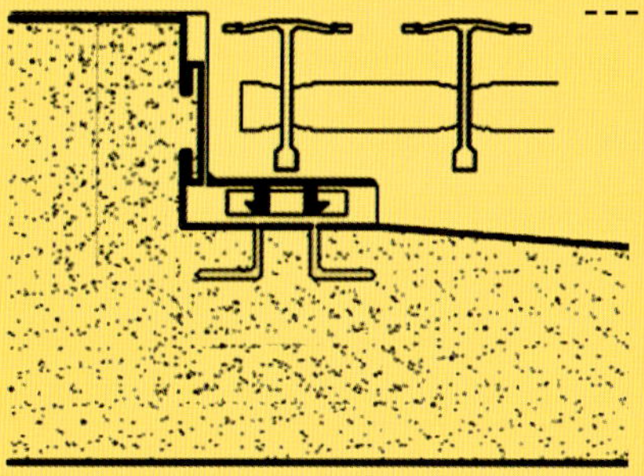

Detail A
详图 A

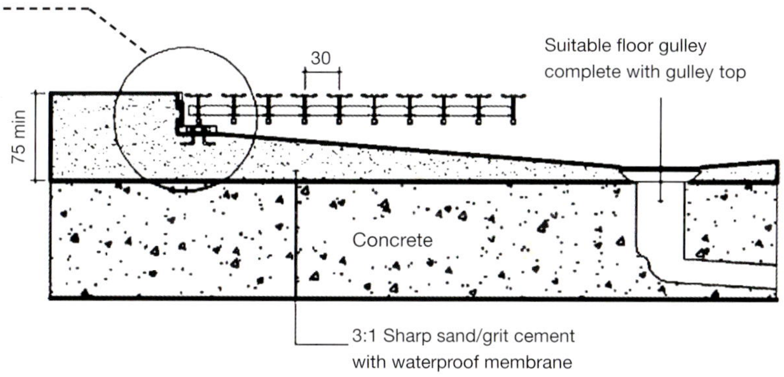

Details of shower floor grills
淋浴地面栏栅详图

章

Cha

pter

7

Renovation and Conversion
整修和改建

7.1 CASE STUDY 1 - TYPICAL CORE A
个案研究 1 - 典型核心筒A

7.1.1 Original Layout
平面图原貌

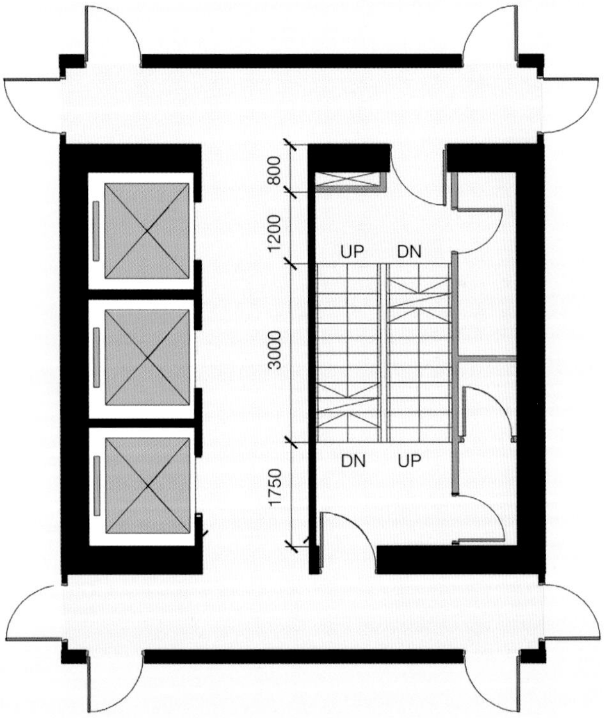

7.1

7.1.2 Approach to Renovation

Core A is a typical core of residential blocks in Hong Kong. In respect of fire safety issues, there is much that could be done to improve accessibility.

Suggested provisions to achieve Universal Design objectives include:

- Accessible refuge area to provide a safe place for the weak and people in wheelchairs in case of fire and emergencies.
- Handrails to assist people with visual impairments or mobility problems.

7.1.2 整修方法

核心筒A是香港住宅楼宇的典型楼层，必须为消防安全考虑更多事宜，以提高畅达水平。

建议可利用下列方法，达到通用设计的目标:

- 设置无障碍避难区，为体弱者及轮椅人士于火警或紧急情况提供安全的避难地方。
- 为视障人士或行动不便人士加设扶手。

7

7.1.3 Enhanced Layout
改良平面图

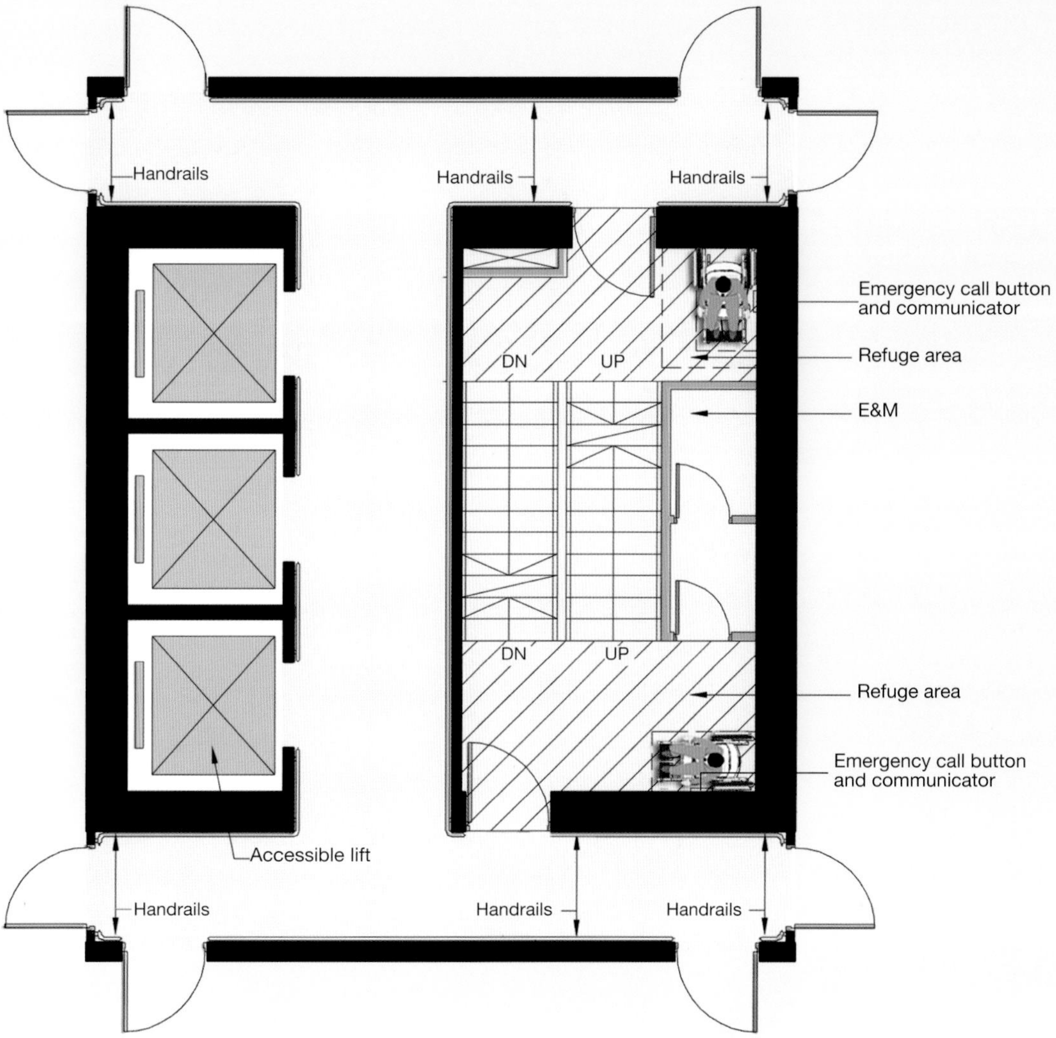

7.2

7.2 CASE STUDY 2 - TYPICAL CORE B
个案研究 2 - 典型核心筒 B

7.2.1 Original Layout
平面图原貌

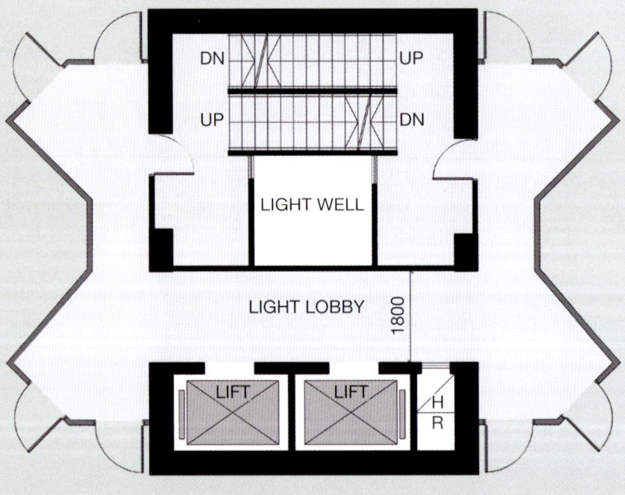

7.2.2 Approach to Renovation

Core B is another typical core of residential blocks in Hong Kong. Suggested provisions to achieve Universal Design objectives include:

- Accessible refuge area to assist the weak and people in wheelchairs.
- Clear escape directions, to assist people with visual impairments.
- Emergency call buttons and communicators at the refuge areas.
- Handrails along corridors, turned up to indicate the location of each entrance, to assist people with visual impairments or mobility problems.
- Braille and raised letter /character labels to indicate flat numbers, direction of lifts and staircases, to assist people with visual impairments.

7.2.2 整修方法

核心筒B是香港住宅楼宇的另一种典型楼层。建议可采用以下方法，以达到通用设计的目标:

- 设置无障碍避难区，以方便体弱者及轮椅人士。
- 清楚显示疏散路线，以方便视障人士。
- 避难区应设置紧急求救按钮及通信器。
- 走廊沿途应加设扶手，并且每户门口向上弯曲以提示入口位置，以方便视障人士或行动不便人士使用。
- 设置凸起盲文和文字/字母的标志，显示楼层及升降机和楼梯的方向，以方便视障人士。

7.2.3 Enhanced Layout
改良平面图

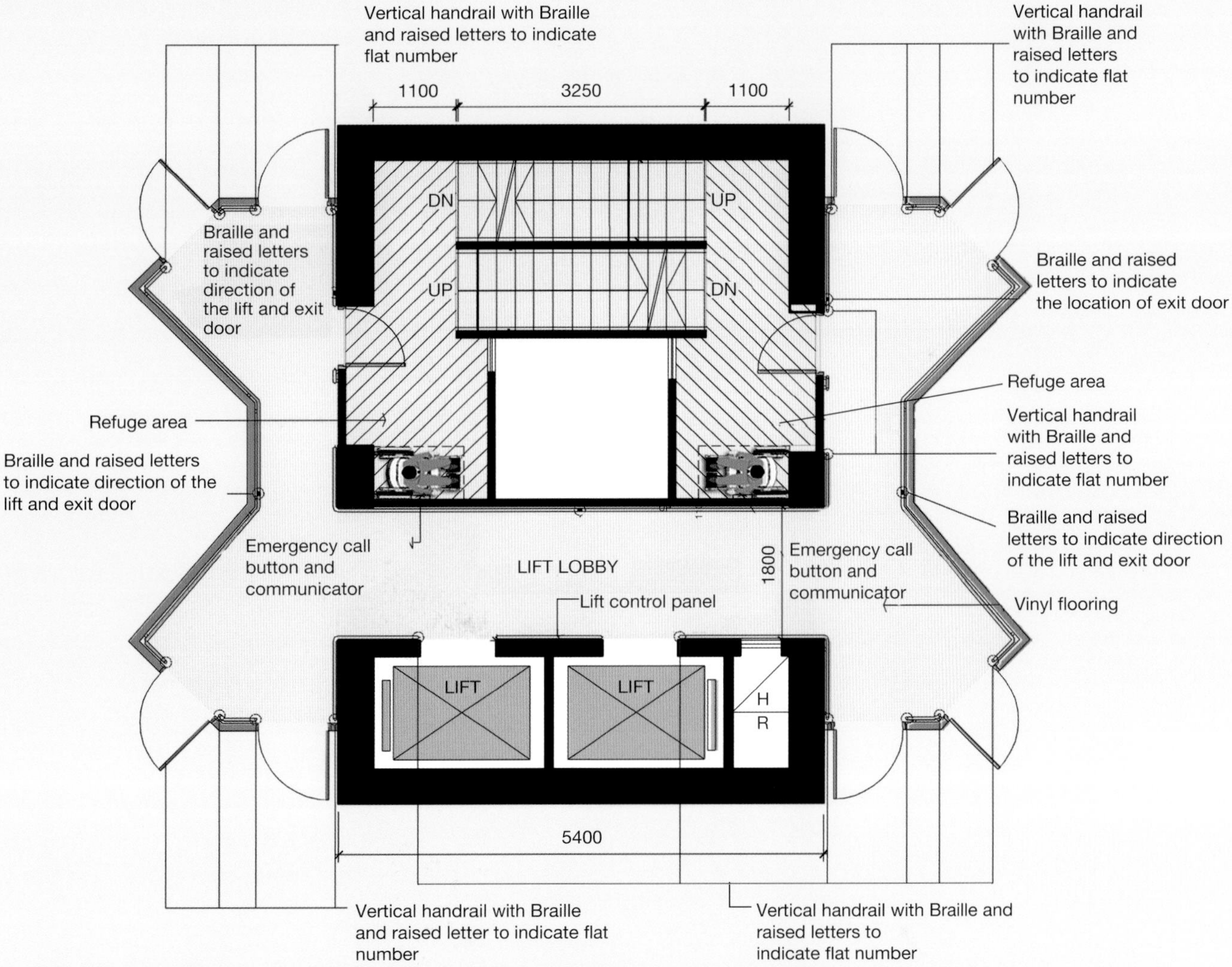

7.3 CASE STUDY 3 - STUDIO FLAT
个案研究 3 - 开放式住宅单位

7.3.1 Original Layout
平面图原貌

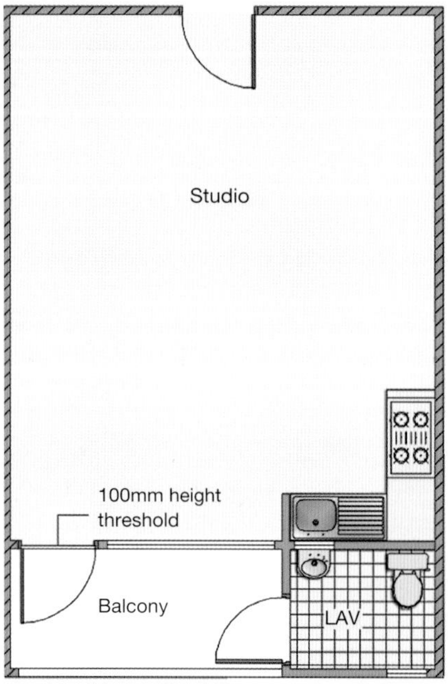

7.3

7.3.2 Approach to Renovation

Suggested changes to achieve Universal Design objectives include:

Bathroom
- Re-location of entrance to Bathroom to improve convenience.
- Replacement of hinged door with sliding door.
- Enlargement of bathroom, providing clear floor space of 800mm x 1500mm, to assist people in wheelchairs.
- Relocation of water closet and provision of a wall-mounted L-shape handrail.
- Provision of knee space under washbasin.
- Provision of a transfer seat to the bathtub.
- Provision of a bath mat.

Pantry
- Enlargement of pantry to include a refrigerator and a washing machine.
- Provision of manoeuvring space to assist people in wheelchairs.
- Provision of seating space in pantry to assist the weak.
- Provision of knee space under the cooktop and sink.
- Provision of a pull-out food preparation top to create working surface with knee space, to assist people in wheelchairs, the elderly and the weak.

7.3.2 整修方法

建议作出以下改变，以达到通用设计的目标：

浴室
- 浴室入口移至室内，方便残障人士。
- 以推拉门替代原本的铰接门。
- 扩宽浴室空间，使地面净空间达800mmx1500mm，方便轮椅人士。
- 抽水马桶置于墙边，加添L形的扶手。
- 在洗面盆下提供膝盖空间。
- 浴缸设置转移座椅。
- 提供浴缸垫。

备餐间
- 扩大面积，以便摆放冰箱和洗衣机。
- 提供操作空间，以方便轮椅人士。
- 在备餐间加设座位空间，以方便体弱者。
- 在炉灶和洗涤盆下提供膝盖空间。
- 设置可以拉出的煮食台，增加使用空间，并预留膝盖空间，以方便轮椅人士、长者和体弱者。

7

Entrance

- Provision of vertical handrail beside door, to assist the weak and the elderly.
- Provision of seating, to assist the weak and elderly.
- Provision of turning space to assist people in wheelchairs.

Balcony

- Replacement of the hinged door with sliding door.
- Removal of the threshold at balcony, to assist people with disabilities.
- Provision of laundry drying space on the balcony.
- Provision of alternative location for washing machine.
- Provision of planters.

入口

- 门边添置垂直扶手，以方便体弱者和长者。
- 加设座椅，以方便体弱者和长者。
- 加设转身的空间，以方便轮椅人士。

露台

- 以推拉门替代铰接门。
- 在露台使用齐平的门槛，以方便残障人士。
- 露台设置晾晒空间。
- 洗衣机可考虑摆放在露台。
- 露台设置花槽。

7.3

7.3.3 Enhanced Layout

改良平面图

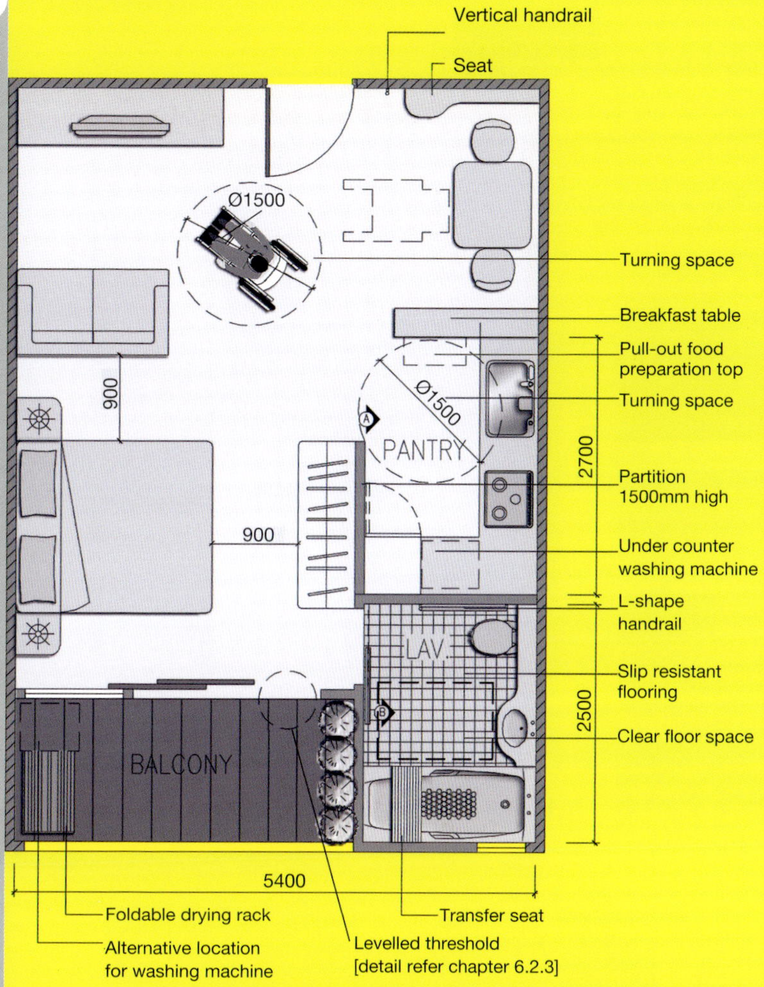

7.3.4 Elevations of Enhanced Layout

改良平面图的立面图

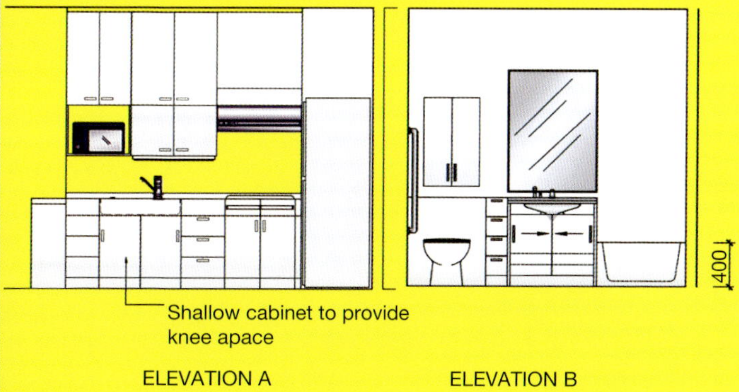

7.4 CASE STUDY 4 - TWO-BED ROOM FLAT
个案研究 4 - 两房单位

7.4

7.4.1 Original Layout
平面图原貌

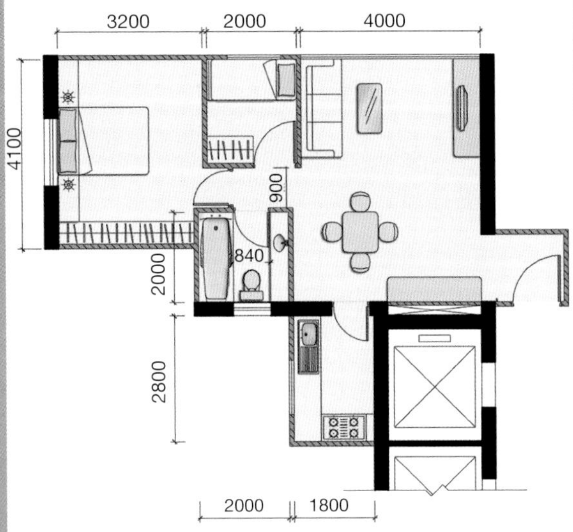

7.4.2 Approach to Renovation

Suggested changes to achieve Universal Design objectives include:

Kitchen
- Provision of manoeuvring space to allow people in wheelchairs to turn 360 degree.
- Provision of seating space, to assist the weak, the elderly and people with disabilities.
- Provision of knee space under the cooktop and sink, to assist people in wheelchairs and accommodate a turning space.
- Provision of a pull-out food preparation top.

Bathroom
- Reduction of bathroom with width of corridor increased to 1200mm to facilitate people in wheelchairs.
- Replacement of hinged door with sliding door, to assist the weak and people in wheelchairs.
- Provision of clear floor area of 1500mm x 800mm for people in wheelchairs.
- Provision of knee space under the washbasin.
- Provision of equipment to assist the elderly to transfer to the bathtub.
- Provision of handrails adjacent to bathtub and water closet, to assist the weak, the elderly and people in wheelchairs.

7.4.2 整修方法

建议作出以下改变，以达到通用设计的目标：

厨房
- 提供轮椅操作空间，方便轮椅人士360°旋转。
- 提供座位空间，以方便体弱者、长者和残障人士使用。
- 在煮食台和洗涤盆下预留膝盖空间，以方便轮椅台人士，并且增加转动的空间。
- 设置可拉出的煮食台。

浴室
- 减少浴室的空间，以便将走廊宽度增至1200mm，以方便轮椅人士。
- 改装铰接门成推拉门，以方便体弱者及轮椅人士。
- 提供1500mmx800mm的地面净空间，以便轮椅人士。
- 洗面盆下设置膝盖空间。
- 提供方便长者可转移至浴缸位置的设施。
- 在浴缸旁边设置扶手，以方便体弱者、长者和轮椅人士。

7

Master Bedroom

- Relocation of hinged door to assist operation by people in wheelchairs.
- Repositioning of furniture to improve accessibility to the window.
- Provision of turning space between the bed and wardrobe, to assist people in wheelchairs.

Small Bedroom

- Relocation of partition adjacent to the living room to increase the width of the small bedroom.
- Repositioning of furniture to accommodate desk and turning space for people in wheelchairs.
- Improvement of access to the window.

Living Room

- Relocation of original cabinet to provide turning space near the entrance, to assist people in wheelchairs.
- Relocation of dining table in the dining area, to assist people in wheelchairs to access the kitchen.
- Provision of a vertical handrail beside the entrance door, to assist the weak and the elderly.

主人房

- 将铰接门移位，以方便轮椅人士使用。
- 重新摆放家具，以方便到达窗户位置。
- 在床铺和衣柜之间预留轮椅旋转空间，以方便轮椅人士。

小卧室

- 向客厅方向移动小卧室隔墙，使小卧室更显宽敞。
- 重新摆放家具，使轮椅人士可使用书桌及增加其转动空间。
- 改变摆设，更容易到达窗户。

客厅

- 重新摆设原有装置，使入口处有更多的轮椅旋转空间，以方便轮椅人士。
- 重新摆设餐桌的位置，使轮椅人士更容易到达厨房。
- 在入口处旁边设置垂直扶手，以方便体弱者和长者使用。

7.4.3 Enhanced Layout
改良平面图

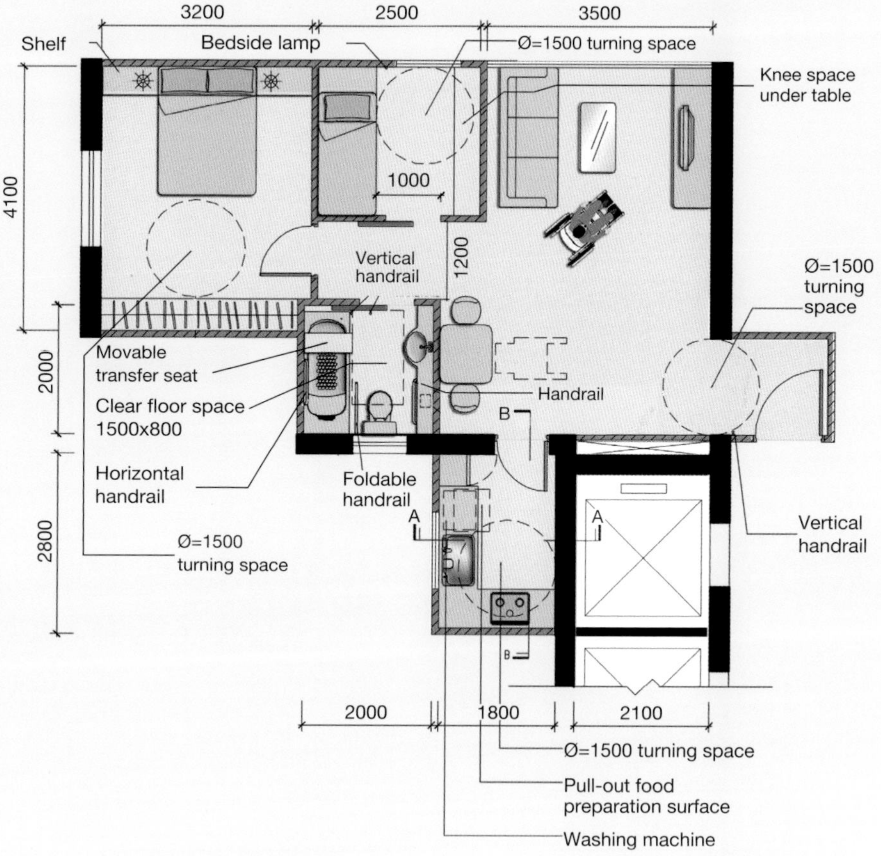

7.4.4 Sections of Enhanced Layout
改良平面图的剖面

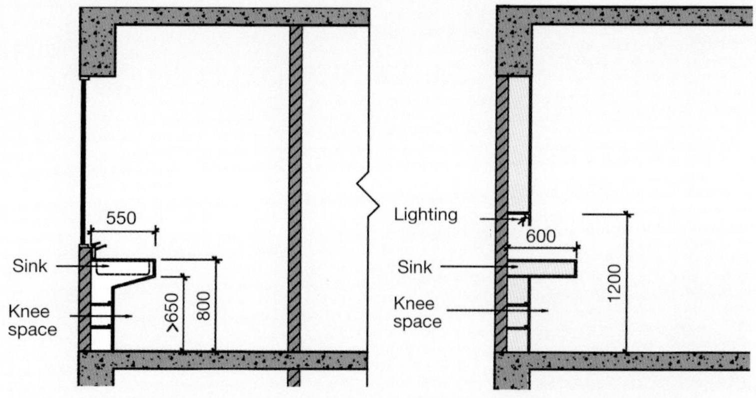

7.5 CASE STUDY 5 - THREE-BED ROOM FLAT
个案研究 5 - 三房单位

7.5.1 Original Layout
平面图原貌

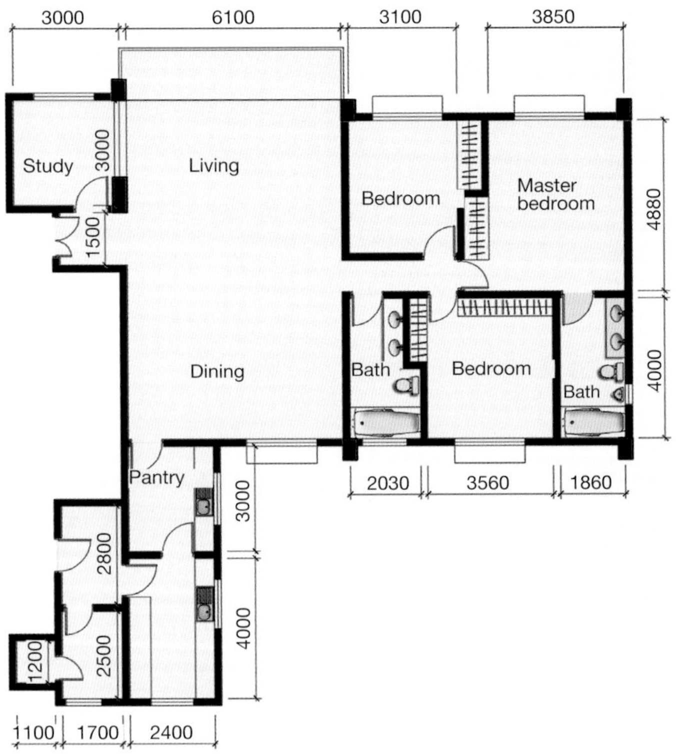

7.5

7.5.2 Approach to Renovation

The original layout has sufficient area to achieve manoeuvring space in all rooms. However, suggested changes to achieve Universal Design objectives include:

Bathrooms

- Provision of handrails in bathrooms.
- Provision of clear floor space of 1500mmx 800mm.
- Provision of transfer seats to bathtubs, to assist the weak, the elderly and people in wheelchairs.
- Provision of tactile positional tiles in front of the urinal in the master bedroom's toilet, to assist people with visual impairments.
- Provision of handrails, to assist the weak, the elderly and people in wheelchairs.

7.5.2 整修方法

原来的平面图已有足够空间在每个房间提供操作空间。但仍作出以下改动，以达到通用设计的目标:

浴室

- 在浴室设置扶手。
- 地面净空间应有1500mmx800mm。
- 浴缸设置转移座椅，以方便长者及轮椅人士使用。
- 在主人房厕所的尿厕前设置方位引路砖，以方便视障人士使用。
- 装置扶手，以方便体弱者、长者和坐轮椅人士使用。

7

Laundry

- Replacement of the original pantry with laundry area not in the original scheme.

Bedrooms

- Reduction of depth of the smallest bed room, with increased width of the corridor to 1200mm, to facilitate access for people in wheelchairs.
- Repositioning of furniture in Master Bedroom to accommodate a caregiver's bed in optional layout.

晒衣处

- 将原本的备餐间改为晒衣处。

卧室

- 减低最小卧室的深度，以扩宽走廊的宽度至1200mm，从而方便轮椅人士进出。
- 重新摆放主人房的家具，在平面另选方案中设置看护卧床。

7.5.3 Enhanced Layout

改良平面图

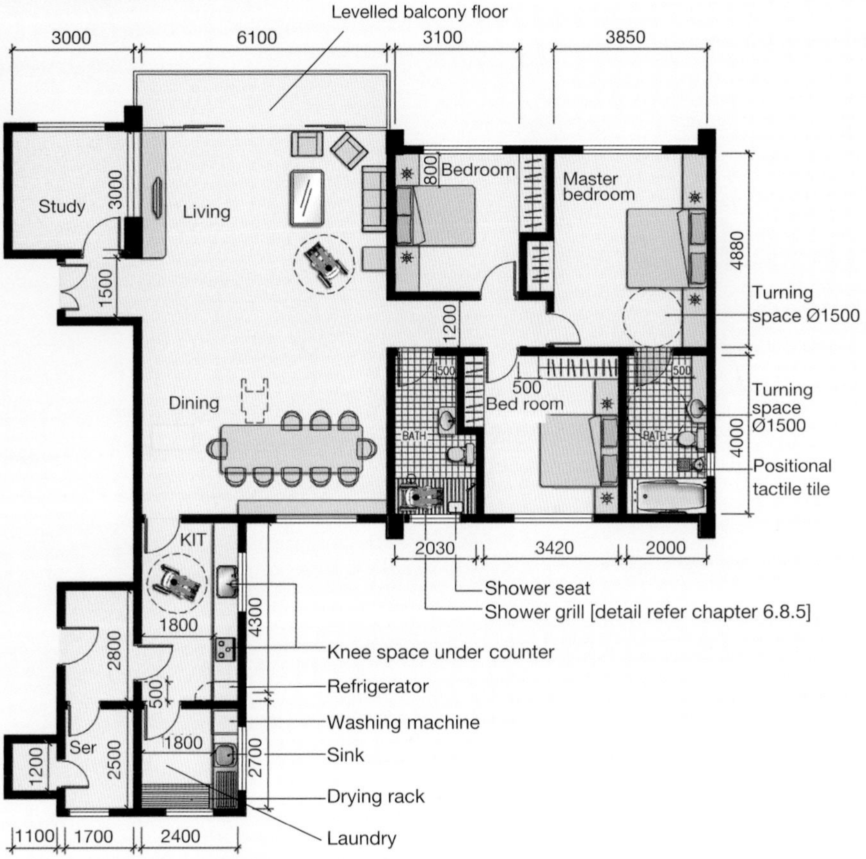

7.5.4 Alternative Layout of Master Bedroom

主人房平面图另选方案

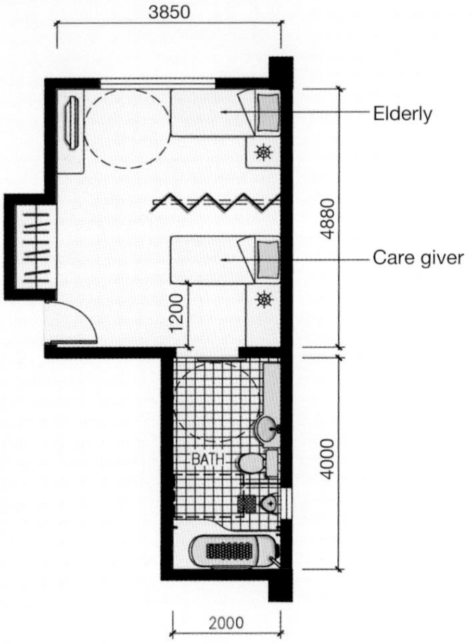

附录I

APPENDIX I

参考书目

BIBLIOGRAPHY

1. American Society of Testing and Materials. Standard Consumer Safety Performance Specification for Playground Equipment for Public Use ASTMF1487-01, ASTM, 100 Barr Harbor Drive, West Conshohocken, PA 19428-2959.
2. The Aging Design Research Program of the American Institute for Architectural Research 1995-96 Trustees. Design for Aging 1996-97 Review. Washington DC: Rockport Publishers, Inc,1997
3. Barrier-Free Design Centre. Architectural Guidelines for Barrier-Free Design. Toronto,1997
4. Bradtmiller, Bruce, Annis James. Anthropometry for Persons with Disabilities: Needs for the Twenty-first Century, Task 2: Analysis and Recommendations
5. Building Authority. Design Manual – Barrier Free Access. Hong Kong: Building Department,1997
6. The Center for Universal Design, North Carolina State University. Universal Design in Housing, North Carolina State University,2003
7. Clara, Yip. A Study on Playground Accessibility in Hong Kong. Hong Kong: Playright Children's Play Association,1999
8. Connell, Bettye; Jones, Mike; Mace, Ron; Mueller, Jim; Mullick, Abir; Ostroff, Elaine; Sanford, Jon; Steinfeld, Ed; Story, Molly; Vanderheiden, Gregg [1997], Principles of Universal design, NC State University, The Center for Universal Design, [Website: http://www.design.ncsu.edu:8120/cud/univ_design/principles/udprinciples.htm
9. Department of Architecture. The Chinese University of Hong Kong, Design Parameters for Elderly Care Architecture in Hong Kong. Hong Kong: The Hong Kong Council of Social Service,2003
10. Dobkin, Irma; Peterson, Mary; [2001], Universal Interior by Design: Gracious Spaces, McGraw-Hill, USA
11. Francis and Francesca Weal. Housing the Elderly – Options and Design. London: Mitchell,1988
12. Glorya Hale. The Source Book for the Disabled – An Illustrated Guide to Easier More Independent Living for Physically Disabled People, their Families and Friends. New York & London: Paddington Press Ltd.,
13. Goldsmith, Selwyn. Designing for the Disabled. Oxford: Architectural press,1997
14. Holmes Shiedle, James. Barrier-Free Design, a manual for building designers and managers. Oxford: Architectural Press,1996

15. Howarth, Peter; Reid, Anita [2000], Allergy-Free Living: How to Create a Healthy, Allergy-Free Home and Lifestyle, Mitchell Beazley, London
16. Kansas State University Agricultural Experiment Station and Cooperative Extension Service. Creating Accessible Homes, Kansas State University,1996
17. Knecht, Barbara. Accessibility Regulations and a Universal Design Philosophy Inspire the Design Process. Website: http://archrecord.construction.com/resources/ conteduc/archives/0401edit-. asp
18. Kwok, YCJ; Siu, KWM. Community Space Research and Design. Hong Kong: The Hong Kong Polytechic University, 2004
19. Levin, B.M. Fire and Life Safety for the Handicapped, National Bureau of Standards. Washington: US Government Printing Office,1980
20. Liauw, Laurence. Elderly Care Residential Typologies. Hong Kong: The Hong Kong Polytechnic University, 2001
21. Lim, Bernard; Leung, MK; Wong, Irene; Chan, Rex; Chung, Edward; Lee, Jocelyn; Giang, Keith; Poon, Viola [2001], Innovative School Design Parameters in Hong Kong for 21st Century, Department of Architecture, The Chinese University of Hong Kong, Hong Kong
22. Mace, Ron. Accessible, Adaptable, and Universal Design. NC State University. The Center for Universal Design,1997
23. Martin Valins. Housing for Elderly People – A Guide for Architects and Clients. Architectural Press,1988
24. Moore, Robin; Goltsman, Susan; Lacofano, Daniel [1987], Play For All - Guideline, Planning, Design & Management of Outdoor Play Settings for All Children, USA
25. Mostaedi, Arian, Residences for the Elderly. Carles Broto & Josep Ma Minguet. Barcelona
26. Preiser, Wolfgang F E . Universal Design Handbook. New York: McGraw-Hill, 2001
27. Raschko, Bettyann. Housing Interiors for the Disabled & Elderly. New York: Van Nostrand Reinhold Company, 1982
28. Regnier, Victor. Design for Assisted Living – Guidelines for Housing the Physically and Mentally Frail. New York: John Wiley & Sons, 2002
29. Regnier, Victor. Design for Assisted Living – Guidelines for Housing the Physically and Mentally Frail. Canada: John Wiley & Sons, 2002
30. ReHabAid Centre, Hospital Authority. Guide on Adapting the Home for People with Alzheimer's Disease – A Manual for Health Professionals. Hong Kong: Hospital Authority, 2003

31. Research Group on Urban Space and Culture, School of Design, The Hong Kong Polytechnic University. Participatory Research - for the Design of Children and Youth Integrated Services Centres. Hong Kong: The Hong Kong Polytechnic University, 2002
32. Rob Davies and Peter Heseltine. A Guide to the European Playground Equipment and Surfacing Standards EN1176: Part1 ~ 7 (Third edition revised) 2000, The Royal Society for the Prevention of Accidents RoSPA, Playground Management Services, Birmingham
33. Robson, David; Nicholson, Anne-Marie; Barker, Neil. Homes for the Third Age – A Design Guide for Extra Care Sheltered Housing, London: E & FN Spon, 1997
34. Steinfeld, Edward; Paquet, Victor. Space Requirements for Wheeled Mobility: An International Workshop, 2004 Website: [http://www.ap.buffalo.edu/idea/space%20workshop]
35. Story, M. Fillette; Mueller, James; Mace, Ronald. The Universal Design File. The Center for Universal Design
36. U.S. Architectural & Transportation Barriers Compliance Board. Americans with Disabilities Act: Accessibility Guidelines for Buildings and Facilities. Washington D.C. : U.S. Architectural & Transportation Barriers Compliance Board
37. U.S. Consumer Products Safety Commission. Handbook for Public Playground Safety, U.S. : CPSC Publication 1997, 325
38. U.S. Department of Justice. Americans with Disabilities Act: Guide for Places of Lodging: Serving Guests Who are Blind or Who Have Low Vision. Washington D.C. : U.S. Department of Justice
39. Working Group on Community Occupational Therapy, Coordinating Committee for Occupational Therapists, Hospital Authority. Guide to Environmental Modification, Occupational Therapy in Practice. Hong Kong: Hong Kong Occupational Therapy Association, 1999
40. UIA, Architects Regional Council Asia. Design for All - An Inclusive Approach. Hong Kong: The Hong Kong Institute of Architects, 2002

41. 日本建筑学会，日本健康环境体系研究会．适应残障者之环境规划．台北：大佳出版社，1986
42. 羌苑，袁逸倩，王家兰．国外老年建筑设计．北京：中国建筑工业出版社，1999
43. 关华山，台湾老人的居住环境．台湾：田园城市文化事业有限公司，1996
44. 胡仁禄，马光．老年居住环境设计．南京：东南大学出版社，1995
45. 中华人民共和国建设部．中华人民共和国国家标准—老年人居住建筑设计标准GB/T50340-2003．北京：中国建筑工业出版社，2003
46. 中华人民共和国建设部．中华人民共和国国家标准—住宅设计规范 GB50096-1999．北京：中国建筑工业出版社，2003
47. 中华人民共和国建设部，中华人民共和国国家标准—城市道路和建筑物无障碍设计规范 JGJ50-2001．北京：中国建筑工业出版社，2001
48. 圣雅各福群会成人服务部家庭生活教育组．平安是福—老人家居安全手册，香港：圣雅各福群会，1985
49. 交通部运输研究所．无障碍交通环境之规划—公共建筑物与活动场所．台湾，1991
50. 内政部社会司委记．生活品质文教基金研究，公共建筑物残障者使用设施文解说之研究，台湾，1991
51. 郭恩慈，古学斌．我们活着依然精彩—让影像诉说长者的日常生活．香港：香港理工大学，2002

附录II

APPENDIX II

DIMENSION TABLE

1. REACH RANGE

1.1 Forwarded or side reach range of children who use wheelchair is recommended as follows:

Age	Reach range
3 to 4 year old	500mm to 900mm
5 to 8 year old	450mm to 1000mm
9 to 12 year old	400mm to 1100mm

尺寸图表

1. 可及范围

1.1 使用轮椅的儿童，其前方和侧边的可及范围如下:

年龄	可及范围
3 ~ 4岁	500 ~ 900mm
5 ~ 8岁	450 ~ 1000mm
9 ~ 12岁	400 ~ 1100mm

1.2 Forward or side reach range of the adult who use wheelchair is recommended as follows:
使用轮椅的成人，其前方和侧边的可及范围如下:

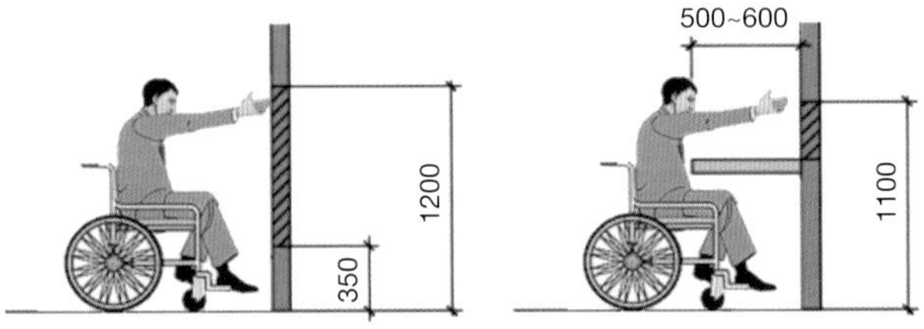

Forward reach
前方可及范围

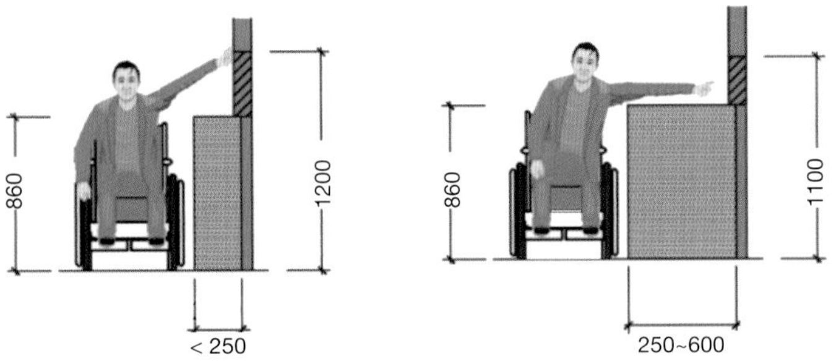

Side reach
侧边可及范围

2. DIMENSION REQUIREMENT FOR GENERAL PATHWAY
一般通道的尺寸规格

2.1 Persons using crutches
使用拐杖人士

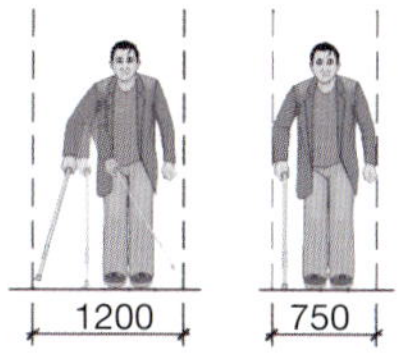

Cane user

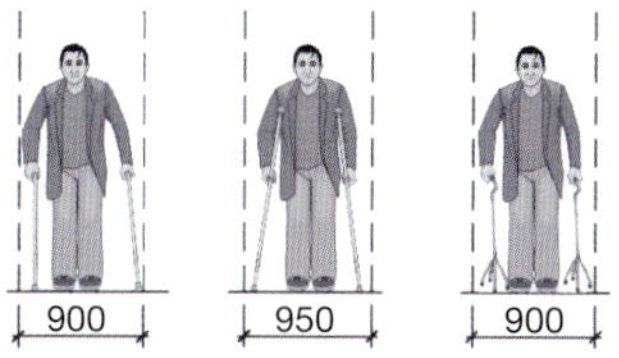

Crutch user

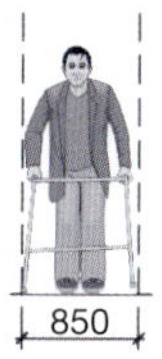

Walking frame user

2.2 Persons with a push chair
使用手推车人士

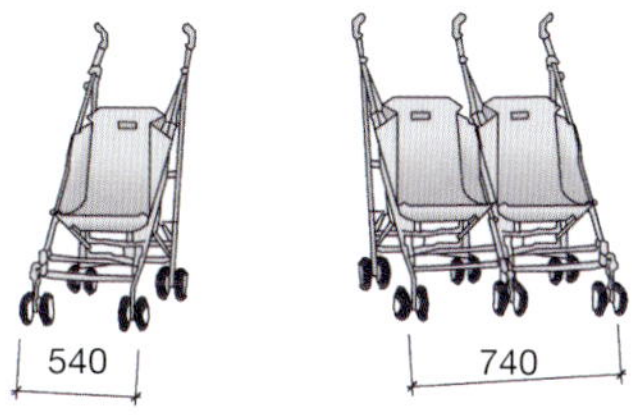

Standard push
chair dimension

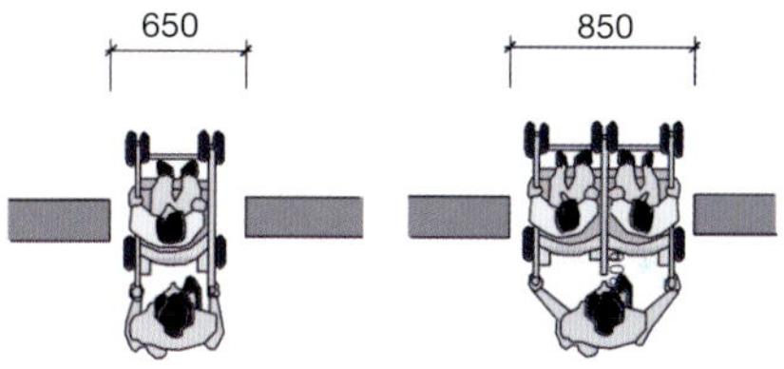

Push chair clear opening
requirements

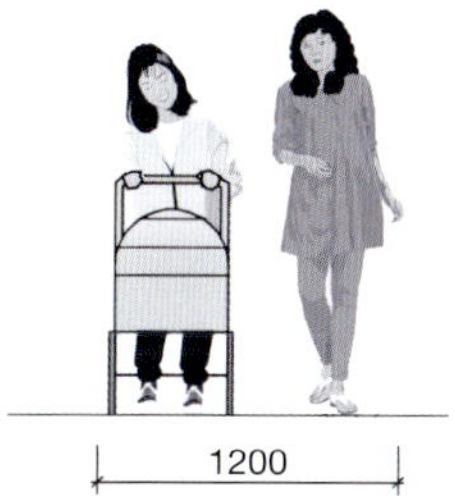

2.3 People in wheelchairs
轮椅人士

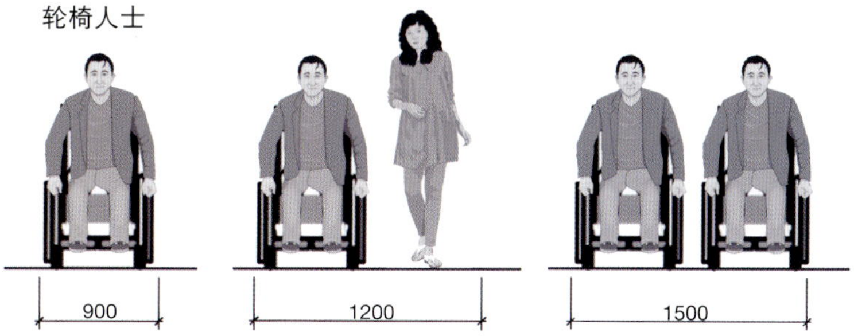

3. DIMENSION OF WHEELCHAIRS

3.1 Comparison of dimension of wheelchairs made in different countries

	Japan	Singapore	America	France
Width (mm)	1040~1050	1100	1065	1200
Length (mm)	610~630	650~700	660	700

3.2 Comparison of dimension of manual and motorized wheelchairs in Hong Kong

	Width(mm)	Length(mm)	Height(mm)	Weight(lbs)
Manual wheelchair	510~725	865~1100	850~1400	23~59 lbs
Motorised wheelchair	520~700	1060~1200	1010~1400	80~220 lbs

3. 轮椅的尺寸

3.1 不同国家轮椅尺寸的比较

	日本	新加坡	美国	法国
宽度(mm)	1040~1050	1100	1065	1200
长度(mm)	610~630	650~700	660	700

3.2 香港手动和电动轮椅尺寸的比较

	宽度(mm)	长度(mm)	高度(mm)	重量(磅)
手动轮椅	510~725	865~1100	850~1400	23~59磅
电动轮椅	520~700	1060~1200	1010~1400	80~220磅

4. Manoeuvring space

4.1 Comparison of requirements on turning circle in different countries

	Japan	Singapore	America	France
90° (mm)	1350 x 1350	1400 x 1400	---	1400 x 1400
180° (mm)	1400 x 1700	---	---	1400 x 1900
360° (mm)	Ø1500	Ø1500	Ø1525	1700 x 1700

4. 操作空间

4.1 不同国家转身空间的规格比较

	日本	新加坡	美国	法国
90° (mm)	1350 x 1350	1400 x 1400	---	1400 x 1400
180° (mm)	1400 x 1700	---	---	1400 x 1900
360° (mm)	Ø1500	Ø1500	Ø1525	1700 x 1700

4.2 Manoeuvring space requirements in Hong Kong
香港操作空间的规格

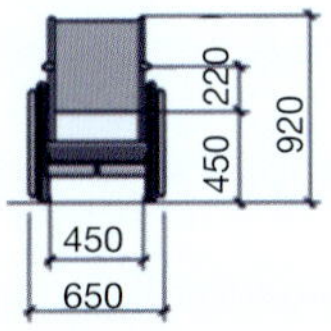

Front

Folded

Rear

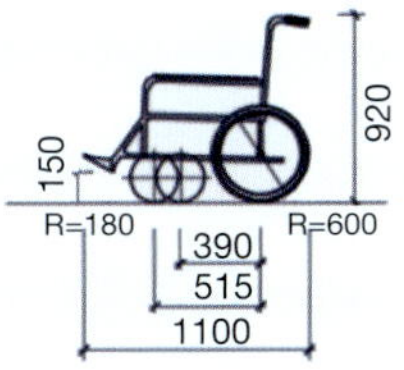

Side view

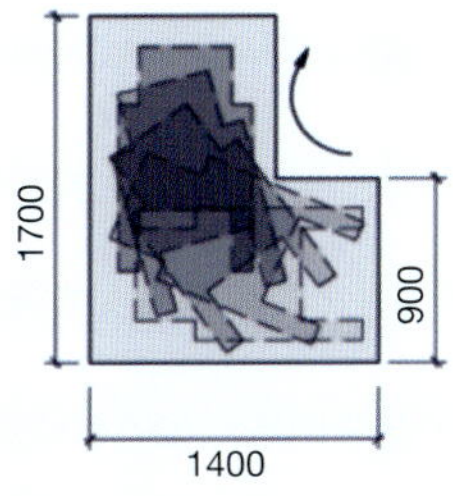

Forward turn through
90 degree

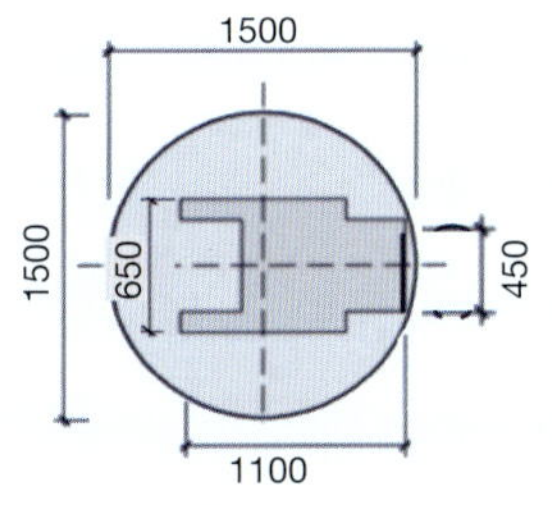

Minimum 180 degree
turning circle

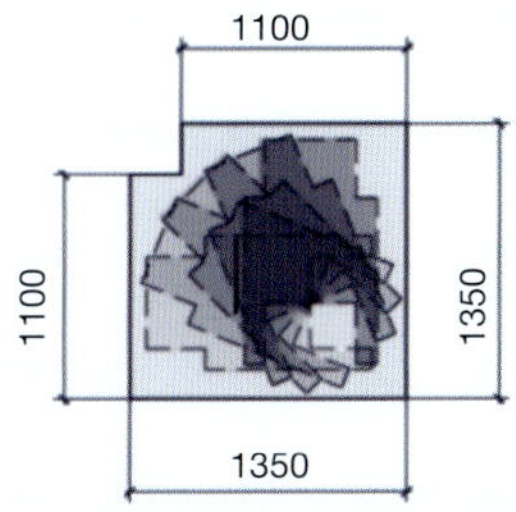

Minimum 90 degree
turning circle

附录III

APPENDIX III

GLOSSARY

In this Guidebook, the following terms are defined as:

1. "Access" is means to enable persons, with or without a disability, to approach, enter or leave a residential building or flat, and to reach and use all facilities of the building or furniture of the flat, without assistance and undue difficulties.
2. "Accessible path" refers to a continuous, unobstructed path that can be accessed by persons with or without disabilities.
3. "Barrier-free design" is an approach to design that removes physical barriers to those with various disabilities.
4. "Cane users" refers to persons who are not able to walk on the level and negotiate suitably graded steps without using canes.
5. "Communal areas" refer to areas inside residential buildings for the common use and benefits of the residents and visitors of the buildings as a whole and not only for the use and benefit of any resident or visitor of any particular unit.
6. "Crutch users" refer to persons who are not able to walk on the level and negotiate suitably graded steps without using crutches.
7. "People with disabilities" refers to persons who on account of injury, disease, or congenital deformity, are impaired in vision, hearing or locomotion.
8. "Refuge area" refers to an area where people may remain temporarily to await further instructions or assistance during emergency evacuation.
9. "Sign" refers to an architectural element composed of displayed text, characters, numbers, symbolic, tactile or pictorial information.
10. "Tactile" refers to an object that can be perceived using the sense of touch.
11. "The people in wheelchairs" refers to persons who, except when using mechanized transport, depend solely on a wheelchair for mobility.
12. "Walking frame users" refers to persons who are not able to walk on the level and negotiate suitably graded steps without using walking frames.
13. "Wheelchair parking space" refers to space to park a single wheelchair.

词汇

以下名词在本指南中的定义:

1. 「畅达」是采用设计手法，使残障或非残障人士进入或离开住宅，且能接触及使用该单位的所有设施及家具并且解决困难，而不需要别人协助。
2. 「无障碍通道」是指连续而无障碍物的通道，可让残障或非残障人士经过。
3. 「无障碍设计」是通过设计手段，为各种残障人士清除实物障碍。
4. 「手杖使用者」是指不能不用手杖在平地行走或爬梯级的人。
5. 「共用地方」是指在住宅建筑物之内，非为某特定单位的居民或访客的利益，而是为居民和访客的整体利益以供共同使用的地方。
6. 「使用拐杖人士」是指不能不用拐杖在平地行走或爬梯级的人。
7. 「残障人士」是指因伤病或天生而引致视觉、听觉或行动能力受损的人。
8. 「避难区」是指在紧急疏散时，供人在其中临时等候，以便获得进一步指示或协助的地方。
9. 「指示牌」是指包含展示性文章、文字、数字、标记、凹凸纹或图象资料的建筑元素。
10. 「凹凸纹」是指可运用触觉去知悉的物体。
11. 「轮椅人士」是指那些非在乘坐机械性交通工具时，只能倚赖轮椅活动的人士。
12. 「步行架使用者」是指不能不用步行架在平地行走或爬梯级的人。
13. 「轮椅停放处」是指可供停放一张轮椅的地方。